Preface

The inspiration for "Ethnobotany and Traditional Knowledge Preservation" arose from a deep recognition of the wisdom embedded in traditional cultures and their enduring relationship with the natural world particularly with plants. Across generations and geographies, communities have cultivated a profound understanding of the flora around them, not merely as resources, but as integral elements of their identity, spirituality, and survival. This knowledge carefully passed down through oral traditions, rituals, and lived experiences is now under threat.

As modern development, climate change, and cultural homogenization accelerate, the urgency to document, protect, and revitalize traditional ecological knowledge has never been greater. Ethnobotany and Traditional Knowledge Preservation is both a scholarly exploration and a heartfelt tribute to those who continue to live in harmony with their environments, guided by inherited wisdom.

Bringing together interdisciplinary perspectives from ethnobotany, anthropology, conservation, and indigenous studies, this volume aims to bridge the gap between scientific inquiry and community experience. We delve into diverse examples of plant use from healing and food sovereignty to spiritual and ceremonial practices and spotlight the stewards of this knowledge: the elders, healers, farmers, and forest dwellers whose voices are too often marginalized.

The chapters that follow offer more than academic insight they offer stories of resilience, innovation, and respect. Our hope is that Ethnobotany and Traditional Knowledge Preservation not only informs but inspires a greater appreciation for the value of traditional knowledge in addressing today's ecological and cultural challenges. In honoring these ancient practices, we also pave the way for a more inclusive and sustainable future.

We extend our deepest gratitude to the knowledge holders, researchers, and communities who shared their time, trust, and teachings. May their wisdom continue to flourish and guide generations to come.

Table of Contents

Dedication

Dedicated to my Beloved Father
Professor (Dr.L.K.Pareek)
I know you would be here if heaven were not so far….

About Author

Dr. Aparna Pareek

Dr Aparna Pareek, MSc, PhD, FBS, UGC Research Awardee and Assistant Professor of Botany at the Department of Botany, University of Rajasthan. Jaipur has a brilliant academic career. She receives the CL Batra Memorial Gold Medal, the Master Amir Chand Memorial Gold Medal at the BSc Examination, and the Gold Medal and Certificate of Merit for First **Position** in M.Sc. Botany Examination. She has pursued her PhD research on the morphogenetic studies in ornamental species of Dianthus. She has twenty-one years of experience in teaching and research, with a focus on Plant Tissue, Organ, and Protoplast Culture. She is currently working on the conservation of Ethnomedicinal plants in Rajasthan that are facing the danger of extinction. She is the recipient of the Certificate of Appreciation for Research from the Department of Science and Technology, Government of Rajasthan, the prestigious APJ Abdul Kalam Award, and the Innovative Botanist Award for her work in the field of Biodiversity conservation. She has authored more than 90 research papers and review articles in Reputable International and National journals. She has actively participated in numerous national and international conferences, seminars, and workshops, presenting research papers. She is the author and editor of nine books in Plant Tissue Culture, Molecular Biology and Plant Morphology and Anatomy. She has been the Principal Investigator of five research projects funded by the UGC, New Delhi and DST, Government of Rajasthan. She is a co-investigator in one project funded by Rashtriya Uchchatar Shiksha Abhiyan (RUSA). She is a Fellow of the Indian Botanical Society, the Mendelian Society of India, and a Life Member of the Indian Science Congress Association, as well as various other academic bodies.

Chapter 1: Introduction to Ethnobotany

In this chapter, you will learn about:

1. **Ethnobotany and scope of the subject area covered by this discipline.**
2. **To understand the concept and importance of Ethnobotany**
3. **Discuss the historical roots of Ethnobotany**

Ethnobotany is primarily the study of the relationship between people and the plants they encounter. Some botanists prefer to define it as the scientific study of the interaction between human cultures and plants. The scope of Ethnobotany encompasses the study of plants used for various economic and social purposes. This research will focus on the contributions of ancient knowledge based on plants to disease cures. The term "Ethno-botany" was given by famous botanist J. W. Hershberger in 1895 to indicate plants used by the aboriginals. "Ethno" is the study of people, and "Botany" is the study of plants. It also assesses the impact of plants on human society in various aspects, including culture, rituals, and healthcare.

The term "Ethnobotany" is not new to India; it was practised here, like many other civilisations worldwide. Ayurveda is an integral part of our heritage, a highly organised ancient science that cures ailments by using plants as medicines. Schultes (1962) defined Ethno-botany as "the study of the relationship between people of primitive societies and plants around them". Earlier, it was based on the relationships and interactions between plants and people in traditional tribal cultures. However, today, this branch of science also includes Western people who utilise plants in modern medicine. Ethno-botanists work to document, describe, explain and validate the knowledge of complex relationships between people and how they use and manage plants as medicines. Ethno-botanists also record the rituals, religious practices, and ceremonial uses of plants in society, discussing the role that a specific plant or set of plants plays within an ethnic group in its natural habitat.

Traditional folk medicine is a broad and vital aspect of Ethnobotany. The traditional art of healing through medicines derived from plants is a significant characteristic feature of Indian culture. The study of the correlation between plants and the people surrounding them is known as Ethnobotany. Some botanists prefer to define it as the research-based (scientific) study of the interaction connecting plants and human cultures. The scope of Ethnobotany encompasses the study of plants used for various economic and social purposes. This research focuses on the contributions of ancient knowledge based on plants to the treatment of diseases. The term "Ethnobotany" was coined by the famous botanist J.W. Harshberger in 1895, who suggested that the aboriginal people had used plants for a long time. Botany is the study of plants, and Ethno means the study of people. It also examines the impact of plants on human society in various aspects, including culture, rituals, and healthcare.

The term "Ethnobotany" is not new to India; it was practised here, like many other civilisations worldwide. Ayurveda is an integral part of our heritage, a highly organised ancient science that cures ailments by using plants as medicines. Schultes (1962) defined Ethno-botany as "The study of the relationship between people of primitive societies and plants around them". Earlier, it was based on the relationships and interactions between plants and people in traditional tribal cultures. However, today, this branch of science also includes Western people who utilise plants in modern medicine. Ethno-botanists work to document, describe, explain and validate the knowledge of complex relationships between people and how they use and manage plants as medicines. Ethnobotanists also record the rituals, religious practices, and ceremonial uses of plants in society, discussing the role that a specific plant or set of plants plays within an ethnic group in its natural habitat. Traditional folk medicine is a broad and critical aspect of Ethnobotany. The traditional art of healing through medicines derived from plants is a significant characteristic feature of Indian culture.

Many well-known workers proposed various definitions of Ethnobotany. According to Jones (1941), "Ethnobotany is the study of the inter-relationship between man and vegetation". Faulks (1958 examined 'The overall relationship between vegetation and man'. Gadgil and Vartak (1976) reviewed the branch of economic botany that deals with the culture of aboriginal or tribal peoples and the role of plants in their lives. Jain (2001) defined it as 'It deals with the traditional inter-relationship between plants and man and his domesticated animals.

There has been a recent boom in Ethnomedicinal plants, primarily driven by interest in traditional herbal medicine (THM), particularly "tribal medicine". The renewed interest in plant-based drugs began because natural medications are considered safer and healthier. Ever since the 29th World Health Organisation assembly in 1976 and 1977 provided official recognition to traditional medicine (TM) and suggested integrating traditional healers (TH) into the rational healthcare programme, these plant-based medicines have proven beneficial during remote times in the Vedic period, approximately 5000-3000 B.C. According to the World Health Organisation, approximately 80 per cent of living beings depend on TM for fundamental health issues.

Ethnobiology is used to research documents and records from ancient times, as well as the knowledge and wisdom of historical people regarding ecological resources. Industrialisation, deforestation, transmigration, colonisation, urbanisation, and more developmental activities (DA) have endangered not only traditional knowledge but also biological resources (BR) and botanical learning (Singh & Pandey, 1998). Hence, preserving and recording the long-lasting folklore and action is essential before useful Ethnobotanical information gets lost. Some of the disciplines related to the subject of Ethnobotany are mentioned here (Fig.).

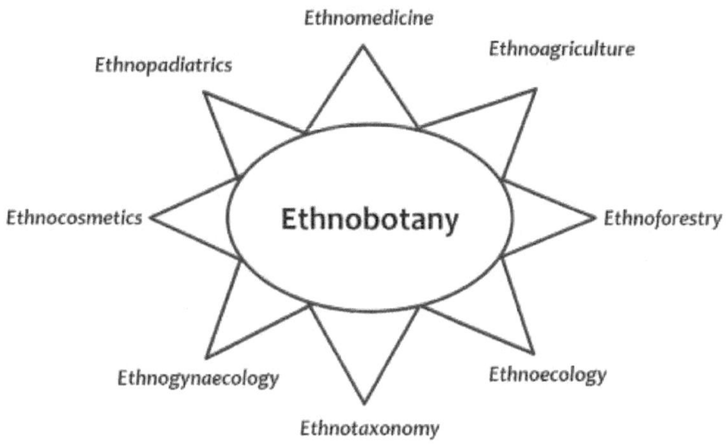

Fig. 1 Subdisciplines of Ethnobotany at a glance

Apart from humans, the environment also harbours a vast plethora of living organisms, comprising plants, animals, and microorganisms, all of which, in conjunction, make this planet Earth a beautiful place to inhabit. Throughout evolutionary history, humans have developed a complex yet compelling relationship with plants, utilising them for various human needs, including nutrition, shelter, clothing, and healthcare. Therefore, it is worth noting that humans would not have progressed as far in the evolutionary timeline without the support of plants. The lack of medicinal health facilities in prehistoric times led to the exploration of plant phytochemicals and their pharmacological properties, which have aided in the sustenance of not only humans but also domesticated animals by curing them of chronic and deadly ailments. This has led to the emergence of two branches of science: "Ethnobotany" and "Ethnopharmacology," which reflect the human dependence on and the mutualistic relationship between plants and humans in meeting healthcare needs.

Ethnobotany is vital in understanding the intricate relationships between cultures and their natural environments. By studying how different communities use plants for food, medicine, and rituals, Ethnobotany preserves traditional knowledge and offers insights into sustainable practices that can inform modern science and conservation efforts. This field emphasises the importance of biodiversity and traditional ecological knowledge in addressing global challenges, including climate change, food safety, and security. Moreover, as many indigenous cultures face threats from globalisation, Ethnobotany helps document and protect their unique heritage, ensuring that valuable plant knowledge is not lost to time.

Ethnobotany is vital for preserving cultural heritage and promoting biodiversity conservation by exploring how various cultures utilise plants for food, medicine, and rituals. Ethnobotany documents traditional knowledge that is often at risk of disappearing. This field not only aids in discovering new medicinal compounds, many of which are derived from plant-based remedies used by Indigenous peoples, but also emphasises the importance of sustainable practices that inform contemporary environmental management. As noted by Martin (1995), "Ethnobotany serves as a bridge between culture and nature," enabling us to appreciate the interconnectedness of human societies and their ecosystems. Understanding these relationships is crucial for addressing and mitigating pressing worldwide challenges, such as food security, climate change, and the significant loss of biodiversity. The main tribes which dwell in the state and utilise these plants for the cure of their various health ailments, as well as cattle, are briefly mentioned -

The impact of Ethnobotanical studies on the conservation of natural resources is direct and significant. On the one hand, it relates to the beliefs, taboos, avoidance, and other constructive approaches of primitive people. On the other hand, the cultivation practices of some primitive peoples, which harm the environment, are highlighted. The aboriginals have preserved the forest through mythological associations as sacred groves. The new areas of observation in the field of Ethnobotany are the selection of species to be consumed, time of collection or exploitation, the conservation through taboos, manipulation of species and habitat protection through faith, processing of wild foods for making them toxin-free, and their preservation, mode of cooking for saving energy and name given to plants. Today, Ethnobotany has become a crucial area of research and development in resource management, biodiversity conservation at the genetic, species, and ecosystem levels, as well as socio-economic upliftment of the region.

Summary

The scope of Ethnobotany has now significantly expanded with the development of new perspectives, both in terms of its theoretical contributions to understanding plant-human relationships and the practical knowledge of tribal people in industries, agriculture, and medicine. The tribal people primarily depend on forest flora to meet their daily needs and basic healthcare requirements. The Ethnobotanical study provides information on the conservation status of plants, including rare, vulnerable, critically endangered, susceptible, endangered, and invasive weeds, which can be used to develop conservation strategies to mitigate habitat loss for essential species. It also opens the door to research into new drugs and chemical compounds found in plants. Several works in Ethnobotany have been conducted worldwide in recent years, highlighting the conservation factors of the most important plants.

Glossary

Ethnobotany: The study of how communities use plants in their surroundings.
Indigenous: Living or occurring naturally in a specific place or environment.

Exercise: Answer the following questions.

1. Define Ethnobotany. Describe the scope of it.
2. Write about the Interdisciplinary approach in Ethnobotany.
3. Explain in detail the conservation of Traditional Knowledge and methods for the protection of Traditional Knowledge.
4. Explain in detail the role of ethnic groups in conserving plant genetic resources.
5. Describe the various methods and techniques used in Ethnobotany for the field-level activities of Data Collection
6. Differentiate between Ethnobotany and Economic botany.
7. Write about the interdisciplinary approach in Ethnobotany.
8. Describe the history and concept of Ethnobotany.
9. Explain in detail the significance of Ethnobotany.
10. Discuss modern Ethnobotany.

Test Paper

Chapter 1
Introduction to Ethnobotany

Time Duration: 30 minutes **Maximum Marks: 25**

General Instructions:
1. The paper consists of 12 questions and is divided into four sections, A, B, C and D.
2. All questions are compulsory.
3. Section A comprises MCQS, consisting of one mark each
4. Section B comprises VSAQS (concise answer questions), carrying three marks each.
5. Section C comprises SAQS (short questions) carrying three marks each.
6. Section D comprises long questions carrying six marks each.

SECTION-A

Exercise 1: Multiple Choice Questions (1*4=4)

1. Who first coined the term Ethnobotany?
(a) Sir Alexander Fleming
(b) John W. Harshberger
(c) C.J. Alexocuplous
(d) J.W. Webster

2. What does the word "Ethno" in Ethnobotany refer to?
(a) Region- Locality- People
(b) Culture
(c) Civilisation
(d) All of the Above

3. The traditional medical system that is acknowledged in India-
(a) Ayurveda
(b) Siddha
(c) Unani
(d) All of the Above

4. Who is known as the "Father of Indian Ethnobotany"?
(a) Dr. S.K. Jain
(b) Dr. R.N. Chopra
(c) Dr. Kaneyala Dave
(d) Dr. Basu

Answer: 1. (b) 2. (a) 3. (d) 4. (a)

SECTION-B

Exercise 2: Very Short answers type questions **(3*3=9)**

1. Define Ethnobotany
2. What are endangered taxa?
3. What are sacred groves?

SECTION-C

Exercise 3: Short answer type questions- **(3*2=6)**

1. Give your opinion about the factors causing biodiversity loss.
2. What are Ethnomedicinal plants?

SECTION-D

Exercise 4: Long answers type questions **(6*1=6)**

1. Explain in detail the conservation of Traditional Knowledge and methods for the protection of Traditional Knowledge.

Online Resources

	Online Practice Paper	Online Video Link
Link	https://forms.gle/wqAk6QmB3k8x3CNm6	https://www.youtube.com/playlist?list=PLzlh_31VWEFrw2yuGOnRKQeZnIcJWoaRP
QR Code		

Chapter 2: Indigenous Communities of Rajasthan

In this chapter, you will learn about:

1. **Various ethnic groups and ethnicities are prevalent in Rajasthan.**
2. **Awareness about the intricate relationship between culture, plants, humans, and our environment.**

The Tribals

Tribal people worldwide have developed their costumes, medicinal practices, and culture based on their local plant environment. Thousands of cultivated and wild plants are being used to help cure various diseases. Therefore, these ethnic groups possess in-depth knowledge regarding the availability and uses of medicinal plants. The plants develop all over them and create an essential form of their culture. Local people rely on medicinal plants, and this traditional knowledge is passed down orally from one generation to the next. Gadialohar, Banjara tribal people are primarily wanderers, moving from place to place within the state and other parts of the country, and they have their own distinct cultural, social, and economic status.

The total tribal population of Rajasthan state is 7,097,706, which accounts for 12.44 per cent of the state's total population, 342,239. The tribes of Rajasthan comprise approximately 8.00% of India's total tribal population (Census, 2001). There is a significant percentage (53%) of Meena tribes in Rajasthan. The literacy rate of tribes in Rajasthan is a mere 10.27%.

Tribal people worldwide have developed their own cultures, costumes, medicinal practices, and more based on their local plant environments. They utilise thousands of wild and cultivated plants to treat various diseases. Therefore, these ethnic groups possess in-depth knowledge regarding the availability and uses of medicinal plants. The Hadoti region is the central tribal belt of Rajasthan, where the main tribes include the Bhil, Meena, Damor, Sahariya, Kalbelia, Kanjar, and Saansi-Nut. The plants growing around them form an integral part of their culture. These people largely depend on their traditional healthcare healing system, and this valuable information is passed on from generation to generation through word of mouth.

Bhil

The Bhil tribe comprises 39% of Rajasthan's total tribal population and is the largest tribe in South Asia. The Bhils are divided into two main groups: the Eastern Bhils, also known as Rajput Bhils, and the Central Bhils, also referred to as Pure Bhils. The Bhils are known as a mountain tribe that uses a bow and arrow. The stories of their exceptional archery talent are recited in the folklore, focusing on Bhil warrior legends. Their primary language is Indo-Aryan, which is merged with the local Hadoti language, commonly referred to as Bhilli. In the history of Rajasthan, the Bhil people are depicted as essential warriors, and many Rajput rulers, such as Maharana Pratap, showed faith in them, including instances during battles with foreign invaders like the Mughals. They also engaged in battles with the Marathas of the Deccan. However, with the changing scenario, they occupied themselves in small-scale agriculture, labour-intensive manufacturing, and repairing machinery. The Bhils are also very good at painting, known as 'Mandno,' which can be seen on the walls of their houses and temples. As forest dwellers, the Bhils possess extensive knowledge of plants used as medicinal herbs.

Meena

The Meena tribe represents about 50 per cent of the tribals in Rajasthan. According to the 'Meena Puran,' the word 'Meena' originated from the word 'fish' and is considered to be the ancestral name of fishermen. However, the remains of Mohan-Jodaro demonstrate the existence of this community prior to the arrival of the Aryans in India. The substances of Meena are closely related to those of the Rajputs. Unlike other tribal communities, the Meenaeena follow a joint family system. The houses of Meena are pretty large, built of mud, timber, and grasses. A group of 10-12 houses forms a 'Dhanni.' Prosperous people, however, construct their houses with stucco, and their main occupation is agriculture, as they have sufficient fertile land. Based on social organisation, they are divided into two main categories:

1. **Jamindar Meena:** They are economically strong and have led a life of privilege in the past. Agriculture, or farming, is the primary occupation, as many landowners in villages near forests are involved in it.
2. **Chaukidar Meena:** The Chaukidar Meena used to have an arrangement with the village people in which they took charges in the form of "Barsi," meaning an annual fee in cash or kind for providing security and safety to a particular group of villages.

Meenas have faith in Hindu mythology and believe in magico-religious and superstitious practices. Liquor and flesh are consumed only on special occasions, such as festivals and ceremonies, but they do not eat fish. They settle marriages outside their clan or subcaste. 'Naata' marriage is socially accepted.

Sahariya

This is the most backwards tribe among all forest dwellers. The Sahariyariya are the only tribe of Rajasthan that the Government of India places in the "primitive tribe" category. Kishanganj and Shahbad tehsils of Baran district are two places where 65% of the total Sahariya tribe population resides. Sahariyas are believed to have originated from the Bhil tribe and mainly reside in the forests of the Baran district. They may have derived their name from the Persian word 'Sehr,' meaning jungle, as they primarily inhabit the jungle and are therefore referred to as Sahariya. They are dependent on farming, fishing, and hunting as their primary source of livelihood.

No genuine historical account of the tribe has been found, but they are believed to be one of the earliest tribes to settle in Rajasthan. They usually live in a separate colony or basti in the village known as Saharana. They are known to be primitive, but some of them also indulged in cultivation and agricultural labour.

Bhil Meena

The Bhil-Meena tribe is believed to be a product of a cross between Bhils and Meenas. Some people consider it a sub-tribe of the Meenas or Bhils. By all standards, they are a section of the Bhil community that, over time, has begun styling themselves as Bhil-Meena to establish their proximity to the Meenas and thereby attain a higher social status in the region. They follow the Hindu religion, and their main occupations are farming and working as agricultural labourers. Their primary language is Marwadi. However, those settled in the Hadoti region speak the Bagri dialect.

Kanjar

This is a stroller and a nomadic tribe. This tribe is primarily located near Kota and is known for its reputation for theft. It is an interesting tradition of putting a few drops of liquor in the mouth of a dead body before burying it in the ground. Before committing a crime, taking the blessings of Hanuman Ji and Chautha-mata is considered auspicious, which is called "Patti Mangana" in their language. The ladies of this tribe are the most beautiful among all the tribes. They are experts in a dance called Chakri.

Sansi and Nut

Sansi and Nut are a semi-nomadic criminal community specialising in hunting and theft. Nuts are considered a subtribe of the Sansi, whose ladies are noted for their good looks. Females are well known for rope dancing, and males are considered for a variety of gymnastic activities. This tribe is found in most of the forest areas near Kota. Bija and Mela are two subcastes among the Sarnati. The tribe is notorious for theft, which has been their primary occupation for a long time; however, they possess excellent medicinal knowledge about the plants found in the surrounding forests.

Damor

The tribe lives in forests and villages of the Hadoti region. The smallest village unit is called a Fala, and the head of the tribe is referred to as a Mukhiya. Men also wear jewellery like women. Like other tribes, they also depended upon forests for food, fodder, and shelter. Their chief food includes 'Khichra,' made from 'Bajara' (Pennisetum americanum) during the winter, which they eat with fried fruits of guar (Cyamopsis tetragonoloba) and Kair (Capparis decidua). They are also fond of liquor. They have faith in 'Hindu mythology. Their 'Bhopa' is called 'Bhagat.' Banjaras are suitable agents for transmitting Ethnobotanical information.

Gadiya Lohar

They are nomadic people, an indigenous community of wandering blacksmiths. They spend their entire lives on the bullock cart, on which the whole household's things are carried, and they settle temporarily on the outskirts of cities. The primary occupation involves repairing iron articles using both hot and cold methods. It is a male-dominated society that places faith in a nuclear family. They settle marriages outside their clan or sub-caste. 'Atta-Satta' marriage is socially accepted in this tribe. This tribe is also renowned for its extensive knowledge of medicinal plants.

Bagaria or Bagri

Bagri is also a criminal nomadic tribe community. They have an efficient court system called a "Panchayat" to resolve their disputes. They have acquired Ethnobotanical knowledge through their nomadic lifestyle.

Garcia

Garasia is a resident of the forest areas of the Rajasthan and Gujarat states. The Garasia is an important tribe in Rajasthan, ranking after the Bhil tribe, and has been derived from 'garas,' meaning 'plenty,' implying the cultivated land they acquired for their livelihood. There are three divisions of Garasia based on their origin: Rajput Garasia, Adiwasi Garasia, and Bhil Garasia.

Agriculture is the primary occupation of this tribe, with maize and pearl millet serving as the staple crops. They live in villages consisting of single-roomed huts, within fields or on hill slopes overlooking the same. The Garasias are renowned for their love for dances and songs, fairs and festivals, and sports. This tribe is famous for leading a more colourful life than other tribes in Rajasthan. The head of the tribe stored various plant materials for medicinal purposes to treat ailments that occurred among the tribal members.

The practice of traditional folk medicine is a vibrant and crucial aspect of Ethnobotany. The traditional art of herbal healing is a characteristic of Indian culture, and tribes have relied on it for ages. These aboriginals use thousands of wild and cultivated plants for the treatment of various ailments. A considerable amount of information on medicinal plants is available to these communities, which needs proper documentation. Rajasthan is one of the largest states and has a rich heritage related to these forests and the aboriginal communities. India has suffered from the loss of its rich heritage due to urbanisation and deforestation. Tribals preserved this knowledge through their conservation-oriented approach, which was passed on from generation to generation through word of mouth. Apart from the main tribes, certain nomadic tribes are also seen utilising the state of Rajasthan and its Ethnomedicinal plants for various health benefits. Indigenous communities and tribes, such as the Kalbelia, Bhils, and Nats, as well as traditional practitioners like Ojhas, Bhopas, and Bhagats, utilise some of these plants. Traditional healers and local tribes play a critical role in preserving the knowledge of medicinal plants in several ways. Traditional practitioners, including Bhagats, Ojhas, and Bhopas, possess a rich understanding of plant-based traditional medicines. People from tribal communities have been utilising Ethnomedicinal plants for a long time to treat various ailments. The traditional knowledge and practices of sustainable harvesting may aid in the preservation of medicinal plant biodiversity and prevent the extinction of plant species in the area. People belonging to tribes such as Kalbelia, Bhils, Raika, Nats, Gadolia-Lohar, Langa, Banjara, and Manganiars possess extensive knowledge of medicinal plants. Therefore, their wisdom and knowledge may aid in the treatment of several ailments as well as the conservation of endangered plants. The people belonging to these local tribes, owing to their cultural and religious beliefs, may aid in the protection of plants, such as in the case of "sacred groves," where tribals preserve plants due to their religious convictions. People belonging to the Kalbelia tribe have developed a unique understanding of the various types of vegetation growing in Rajasthan and utilise them as natural and herbal remedies, as well as an alternative source of income. In addition to Kalbelias, the Raika is an indigenous community that has inhabited the state of Rajasthan for over 700 years and acts as custodians of the environment. The Raikas have developed a variety of livestock breeds based on their ancient and traditional knowledge, utilising communal lands and forests for animal grazing. The Raikas have faced and dealt with several natural calamities, including fighting forest fires, tackling the emergence of invasive species, and reporting illegal logging and animal poaching to protect the forests. The grazing pattern of Raikas is unique and stimulates plant growth, as their grazing patterns are based on traditional ancient knowledge and involve a strict rotation over a five-year period. Therefore, both the Kalbelia and Raika tribes have formed a close connection with the natural environment, as well as with the local ecosystems in Rajasthan. Furthermore, their traditional knowledge, use of sustainable practices, and custodians of the environment play a crucial role in preserving medicinal plants and further contribute to the enrichment of biodiversity in the region.

Summary

The ethnic people, mostly indigenous tribals, live near forests and have managed and conserved the biodiversity of their localities for a long time. Sacred groves, taboos, sacred plants, and tree worship culture are excellent examples of the conservation patterns of flora by ethnic groups. Our old literature is a good source of knowledge about the traditional uses of plants. The old literature primarily includes works related to Vedic and Ayurvedic traditions. Ayurveda is a 5,000-year-old system of natural healing.

Glossary

1. Ethnicity: A cultural and social identity of an ethnic group.

2. Ethnobotany: It is the study of how people of a particular culture and region use indigenous plants.

3. Ethnolinguistic group: A group of people that is unified by both a common ethnicity and language.

Exercise: Answer the following questions.

1. What is Ethnobotany?
2. Describe the different tools and techniques of Ethnobotanical studies.
3. Describe the contribution of tribals in the conservation of the forest.
4. What is an ethnic group and ethnicity?
5. Where is the Meena tribe found in Rajasthan?
6. What is the chief occupation of the Saharia Tribe in Rajasthan?
7. What is the difference between tribal and indigenous people?
8. What is the relation between tribal people and plants?
9. What are the Nomadic Tribes?
10. Where are Gadia-Luhar found in Rajasthan?

Test Paper

Chapter 2
Indigenous Communities of Rajasthan

Time Duration: 30 minutes **Maximum Marks: 25**

General Instructions:
1. The paper consists of 12 questions and is divided into four sections, A, B, C and D.
2. All questions are compulsory.
3. Section A comprises MCQS, carrying one mark each.
4. Section B comprises VSAQS (concise answer questions), carrying three marks each.
5. Section C comprises SAQs (short questions) carrying three marks each.
6. Section D comprises long questions carrying six marks each.

SECTION-A

Exercise 1: Multiple choice questions: **(1*4=4)**

1. The book 'Dictionary of Indian Folk Medicine and Ethnobotany" was written by:
(a) G. R. Rao
(b) S. K. Jain
(c) P. K Gupta
(d) Hashberger
2. Sacred Groves are:
(a) Parts of forests that have been left untouched by the local people.
(b) Places for grazing animals.
(c) Forests earmarked for the commercial felling of trees.
(d) Forests are used for planting trees with medicinal properties.
3. The IUCN Red List of threatened species was founded in
(a) 1946
(b) 1948
(c) 1964
(d) None of the above
4. Bhils are found in
(a) Rajasthan
(b) Bihar
(c) Punjab
(d) Delhi

Answer: 1. (b) 2. (a) 3. (c) 4. (a)

SECTION-B

Exercise 2: Very Short answers type questions **(3*3=9)**

1. What are the main tribed found in Rajasthan?
2. Which tribe is the largest in Rajasthan?
3. In which districts of Rajasthan are Bhils prominent?

SECTION-C

Exercise 3: Short answer type questions. **(3*2=6)**

1. What are the traditional occupations of tribes in Rajasthan?
2. What is meant by indigenous communities?

SECTION-D

Exercise 4: Long answer type questions. **(6*1=6)**

1. What is the role of indigenous people in the conservation of Plants?

Online Resources

	Online Practice Paper	Online Video Link
Link	https://forms.gle/X6kaaqwTZXzF6VYY8	https://www.youtube.com/playlist?list=PLzlh_31VWEFrzAqQgWXforaGcQF_Dbx7Q
QR Code		

Chapter 3: South East Rajasthan: Geography, Climate, and Vegetation

In this chapter, you will learn about:

The geography, climatic conditions, and vegetation of the south-east part of Rajasthan, where tribals are found.

The state of Rajasthan is an abode for Indigenous people who are using these medicinally important plants for the sure of their health ailments. Primarily, the tribals reside in the southeastern region of Rajasthan; therefore, the topography and vegetation of this region are described in detail. The Hadoti Plateau, with a rich floristic diversity, is situated at the edge of the Malwa Plateau. The region boasts one of the best scopes of genetic resources for future development in the peninsula, thanks to its flourishing biodiversity. It possesses great potential in the form of a gene pool due to its wide range of Flora and fauna. The Hadoti region's contribution to regulating climate and rainfall patterns is significant. The historical and cultural heritage, as well as its geographical location and physiographic features, make this plateau a unique location.

A Perennial River, the Chambal, drains a large part of the study area. This region is locally known as Pathar (stony) and Uparmal (high tableland) due to the distinct variation in landforms. In Rajasthan, this plateau lies in the eastern parts, along the Chambal River, and covers the greater part of the Bundi, Kota, Baran, and Jhalawar districts. It comprises 9.6% of the Area of the whole of Rajasthan. The tableland has a very diverse topography, consisting of more or less sandy uplands, broad depressions, and level stretches of deep, black Soil. Most of the plateau has a general slope towards Gwalior and the catchment area of the River Betwa. Most of the parts of this region are drained by the Chambal River and its tributaries, including Kali-Sindh, Parwan, Aahu, Ujar, Parwati, and Chandrabhaga. This Hadoti plateau is further divided into two subregions (i.e. in Vindhyan Scraps and Deccan Lava Plateau) based on the regional variations of characteristics.

Vindhyan Scarp Land

The exact extent of the Vindhyas is not clearly defined because it does not form a single mountain range in the geological sense. Hence, it is described as a mountain range composed of several distinct hill systems located throughout the region.

The scarped lands formed by massive sandstones separated by shale resulted from the heavy pressure exerted on them by the Earth's core during its evolution. This part is characterised by an average elevation between 360m and 590m. The scarps face towards the south and southeast, near Ramganj-Mandi and Digod, and extend into the western part of Madhya Pradesh. It represents a unique topography characterised by boulders, low hills, blocks, and depressions. The Vindhyan Scarps are formed from sandstone over a long period of evolution in this Area. Many important tributaries of the Ganga-Yamuna system originate from the hills of the Vindhyas, which makes this Area geographically important and highly fertile. Rivers, including the Chambal, Betwa, Kali Sindh, Aahu, and Parban, drained most of the northern slopes of the Vindhyas.

Deccan Lava Plateau

These 'Uparmal' or high hills represent the Deccan Lava Plateau. It is a broad stony highland, including a triangular alluvial basin made by the rivers Chambal, Kali-sindh and Parbati in Jhalawar district.

Structurally, the Vindhyan group of rocks is somewhat more stable, but some of them are less stable over a greater part of the Area; they exhibit steep, deep gorges. The Central Hadoti Plateau is part of the central drainage basin of the River Chambal, which is a northward extension of the Malwa Plateau, known as the Hadaoti Plateau. Topographic variations, including flat landmasses, Hills, and gorges, characterise the plateau. However, Central Hadoti is mostly a riverine tract with an average elevation of 320 meters above sea level. The southeastern part of the region is a hilly area and forms part of the Mukandara hills.

The Mukandara hills run across the southwestern part of the Central Hadoti. It is a characteristic feature of the Hadaoti Plateau, consisting of a continuous chain of hills known as the Aravali range, which extends from Bundi to Mukandara. The Mukandara hills are a critical formation of two separate parallel ridges, with an intervening gap of 160 metres wide, covered in thick forest. These hills traverse the region for about 145 kilometres. This topographic feature offers a beautiful scenic view of the Area. Darrah Sanctuary is located in the hilly tract zone of the Mukundara Hills. These are extended in the form of a series, covering an area of approximately 6022 sq. km. These series comprise the upper Vindhyan strata of limestone, with various branches present within this series. The hilly tract of Bundi and the western zone is composed of Aravalli rocks. Two prominent hilly tracts in Hadoti, namely Bundi and Mukundara, provide a unique topography to this region. Mukundara ranges extend up to 120 km in the central part of the Hadoti plateau. These ranges enter the Jhalawar district from the southeast side, and two ridges pass close to Jhalarapatan. Both ridges are at an approximate distance of 150 m from each other. These ranges provide ideal environmental conditions for the development of dense forests and flourishing vegetation. Many small rivulets make their way through these ranges during the monsoon. The famous "Darra" pass is a distinguishing feature of the Hadoti region. The highest peak of the Aravali range in this Area is located in Jhalawar district, approximately 490 meters high, near Ravtha village. A peak of 450 meters high is also nearby, near Narayanpura panchayat. Some other central high hills are located in the Ramganj-Mandi tehsil of Kota district, with isolated peaks also found near Pipalda and Indergarh.

The rest of the region is drained by the rivers Chambal, Aahu, Kalisindh, Parbati and other small rivers. This region is a fertile plain, beneficial for agriculture and other economic activities due to the presence of rivers and the silt they bring.

The Hadoti region is characterised by the most fertile plains in the region and sharp topographical disturbances in the form of high hills, a discreetly distributed plateau, and some lower hilly areas. The complete Area is an extension of the upper Vindhyan ranges. The central region of the Area is traversed by the Chambal River and its tributaries, surrounded by hills. The southern part of the Hadoti region features a plateau surface, whereas the southwestern and eastern parts exhibit upland topography. Although the general slope of the Area runs from southwest to northeast, in Gangdhar tehsil in Jhalawar district, it is from southeast to northwest. Based on physiography, altitude, and topography, the Hadoti plateau is divided into five specific zones to study and thoroughly analyse the floristic diversity and vegetation of the Area.

- Hilly tracts of Bundi and Mukandaras.

- The riverine zone along with Jawahar Sagar Dam.
- Shahabad upland with the presence of horseshoe-shaped hilly tracts (Kunda Khoh, Madho Khoh, Sukha Khoh, etc.)
- Jhalawar plateau and hillocks of Jhirniya, Manoharthana and Baghel Ghati.
- Dag-Gangadhar upland (the smallest physical unit of Area)

Type of Soil

Soil is one of the most important natural resources because humanity has depended on Soil for growing crops and providing food. Soil formation is the result of the interplay between the climatic forces and the geomorphic agents. Soil-forming processes determine the nature of available Soil and regulate it. The soils here have developed under harsh, arid, and humid climates. The tints of Soil, chemical composition, texture, and structure of soil particles show large-scale variations under different environmental conditions. The dominant type of Soil is alluvial due to the presence of the perennial river Chambal and its tributaries in this Area. Rivers like Chambal, Kalisindh, Parbati and their tributaries bring this fertile Soil with them. This type of Soil is generally considered suitable for productivity and produces a wide range of crops. As mentioned earlier, this region is a northern extension of the Malwa Plateau. Consequently, black Soil is found in its southern parts, while reddish-brown Soil is the dominant type in the Sultanpur, Ladpura, Barod, and Degod areas.

Southeast Rajasthan is related to the Deccan lava plateau; therefore, predominantly black cotton-growing soils are significantly recorded in this Area. The soils in the Pathari (Stony) area of Jhalawar are primarily formed by the disintegration of the Ganugarh Shales, Samariya Shales, and the Jhiri Shales, intermixed with disintegrated sandstone grains. All over the forest areas present in this hilly region, the Soil is generally dry, impoverished and deficient in humus. On the slopes of hills, the Soil is sandy-to-sandy loam, shallow and covered with stones and boulders. The Soil on the plateaus and other flat areas is sandy loam, derived from the parent rock. In depressions of hills along the banks of streams and low-lying areas, the Soil is clayey loam and is pretty deep.

Black Soil: The Soil of the Kota district and nearby river-drained areas is described as deep, medium, and black alluvium. The Soil ranges in depth from shallow to very deep and is characterised by lime and gravel. The soil type is generally clay loam to clay in texture. Soil develops cracks during the dry season and becomes moderately to less permeable. The stratigraphic succession in the district is as follows: Quaternary, Recent to sub-recent, Alluvium, Clay, silts, sand, kankar, and gravel with rock fragments. The Soil in the Hadoti region is fairly good, but this type of Soil and vegetation vary with the availability of water. The presence of vertical cracks of varying sizes in soil particles is fairly good, and they absorb water well when available.

Components of Soil

Soils of Southeast Rajasthan are a result of the combined interaction of living organisms and climate on barren rocks. Here, we discuss some important nutrients of Soil to understand its nature and impact on the vegetation of this region.

1. Organic carbon: Organic matter contributes to enhancing the nutrient turnover and carbon exchange capacity of Soil, as well as improving soil structure and moisture retention. The range of % OC is 0.51% to 0.85%.

2. PH: It mainly affects plant nutrient availability; the pH range varies at different locations from 7.34 to 8.24.
3. Electrical Conductivity: Soil electrical conductivity is a measurement of various parameters that correlate with different properties found in Soil. The range of Electrical conductivity is 0.42 to 0.70 dS/m.
4. Copper: Copper plays a crucial role in seed production and chlorophyll formation in plants. The range of Cu is 66.56 to 86.55 ppm.
5. Iron: Iron is crucial for producing chlorophyll; hence, its deficiency causes chlorosis. The range of iron is 33.54 to 54.12 ppm.
6. Zinc-Zinc: It is an essential micronutrient for plants to grow well under good physiological conditions. The range of Zn is 0.54 to 0.85 ppm.
7. Manganese: Mn plays several roles and is important in photosynthesis, the synthesis of chlorophyll and nitrogen absorption. The range of Mn is 14.52 to 17.11 ppm.
8. Nitrogen (N): Nitrogen is one of the three primary parameters of soil analysis and is crucial for the overall growth of plants. The range of N is 0.54% to 0.95%.
9. Phosphorus (P): Phosphorus is a crucial soil micronutrient that enhances a plant's resistance to diseases. Phosphorus was found in the range of 21.45 to 42.85 Kg/ha.
10. Potassium (K): It is an important nutrient for various physiological processes such as photosynthesis; potassium is found in the range of 404.54 to 642.82 Kg/ha.

Forest Area

The forest cover in Southeast Rajasthan is 116,500 hectares, which accounts for 22.40% of the total Area. The primary forest area is located in the southwestern part, specifically in the Mukandara hills, which fall within the Jhalawar district. Other dense forest areas, rich in biodiversity, are present in the Shahabad tehsil of Baran district and the Rameshwar Mahadev forest of Bundi district. These areas depict the living heritage of Hadoti. The floristic diversity of any given area is determined by its geographical location, geological history and its past connection with contiguous areas. Taking the above facts into consideration, it is concluded that the vegetation of southeastern Rajasthan generally reflects a mixed deciduous forest type, with a clear indication of the adverse effects of climatic, edaphic, and biotic factors on it.

The forests of the region under study fall under Tropical Dry Deciduous Forest, which can be further divided into three sub-categories:

1. Anogeissus pendula Forest:

In this forest, deciduous trees exhibit luxuriant growth in the dense forests that thrive on hilltops. The forest of the Area also exhibits altitudinal zonation of the vegetation, where Anogeissus pendula is the dominant species, extending from the base to the top of the hills.
The main species found in the region include Anogeissus species, Lannea coromandelica, Aegle marmelos, Acacia leucophloea, Boswellia serrata, Diospyros melanoxylon, and Butea monosperma, among others, in various zones.

2. Miscellaneous Forest:

Kishanganj and Manohar Thana tehsils are hilly terrains and support mixed deciduous vegetation. Tectona grandis is a significant tree species that exhibits biodiversity alongside other tree species, such as those of the Anogeissus pendula type, in the forest. In the peripheral region, the destruction of forests has led to the growth of scrub jungles, including Acacia catechu,

Acacia nilotica, Acacia leucophloea, Belanites aegyptiaca, Terminalia Arjuna, and Terminalia bellerica.

3. Subtropical Evergreen Forests:
The Baran district near Bhanwargarh and Sitabari represents a typical subtropical evergreen forest. These dense Forests harbour the growth of Bombax ceiba, Cordia dichotoma, Mangifera indica, Emblica officinalis, Cassia fistula, and Madhuca indica, among others.
The riverine zone of the Chambal Valley exhibits luxuriant vegetation dominated by Syzygium cumini, Dendrocalamus strictus, Terminalia Arjuna, and other evergreen tree species, as well as some ferns and herbs, such as Boerrhavia diffusa and Convolvulus microphyllus, near water streams.

4. Babul (Acacia nilotica and species) Forest:
Babul and Khair are the representatives of the vegetation found in wastelands. It is mainly found in Chambal, Bihar, and is also present in mining areas such as Dabi, Budhpura, Suket, Arolia, etc.

Apart from the above forest areas, some thick natural forests are in sheltered patches, typically restricted to different wildlife sanctuaries and national parks.
1. Bhensrodgarh Wildlife Sanctuary, Kota.
2. Sorsan Wildlife Sanctuary, Baran.
3. Ramgarh Vishdhari Sanctuary, Bundi.
4. Shergarh Sanctuary, Baran.
5. Mukundara Tiger Reserve, Jhalawar.
6. Mukundara Hills National Park, Jhalawar.
7. Jawahar Sagar Wildlife Sanctuary, Kota.
(Source: Census of India: Administrative Atlas, 2001; Bhalla (2016)

Climate and Rainfall
The Hadoti region experiences a very humid climate, which can be classified as semi-arid. The summers are hot and dry, and the winters are cold. The cold season prevails from December to February, followed by the hot season from March to June. After the summer, the rainy season begins with the onset of monsoon rains, which last until the end of September. The monsoon period is characterised by bright sunlight and temperatures ranging from 35 to 40 degrees. January is the coldest month, with daily maximum temperatures ranging from 18°C to 23°C and daily minimum temperatures ranging from 8°C to 10 °C. The daily maximum temperature during summer ranges between 40 °C and 44 °C, and the daily minimum temperature is between 24 °C and 28 °C, influenced by high temperatures in Kota, Baran, Jhalawar, and Bundi. Summers are very hot, whereas winters are cold and dry. Most of the rainfall occurs during the rainy season, which typically spans from July to September. A wide range of variations in rainfall is found from year to year. Every second or third year is usually a low rainfall year. On average, there are about 45 rainy days, and the annual rainfall varies from 600mm to 990mm. Generally, the Monsoon starts at the beginning of July but is sometimes delayed until the middle of July and continues until the end of September. The rains also occur occasionally during the winter months of January and February and are known as "Mawath".

Rivers

The Chambal is the principal river of the region, and it is the only perennial river of Rajasthan. It originates from Madhya Pradesh, near Mahu, and enters the Kota district on its western border at the tri-junction of the boundaries of Kota, Bundi, and Chittorgarh districts. Kalisindh also meets the Aahu River in front of the Famous Gagron Fort.

The Parwati River is the main tributary of the Chambal River. After originating and flowing through Madhya Pradesh, it enters the Baran district, forming the northeastern boundary between Kota and the state of Madhya Pradesh. Many other distributaries flow in this region, including Aahu, Parwan, Ujar, and Ghoda Pachad, among others. However, all of them are seasonal, with the majority flowing primarily during the monsoon season.

Summary

Rajasthan is situated in India's northwestern region, making it more vulnerable to climatic fluctuations. Rajasthan's western region is barren and dry, whereas its southwest is hilly and moist. Rajasthan's varied climate is evident throughout the state. Desert regions typically have frigid winters and scorching, dry summers. As you approach the Aravali Mountains in the west, the climate, rainfall and humidity are high. At the same time, the climate to the east is more conducive to better rains and heavy humidity. Learning about the climatic conditions of Rajasthan helps us better understand the state's geographical conditions.

Glossary

1. Seasons Meteorological seasons over India are: **Winter Season:** January – February, **Pre-Monsoon Season: March, Southwest Monsoon Season: June - September, Post-Monsoon Season:** October – December. Monsoon: The seasonal reversal of winds and the associated rainfall. This word is derived from the Arabic word "Muslim".

2. Temperature: A body's temperature determines its ability to communicate heat to other bodies or to receive heat from them. Meteorologists are interested in the temperatures of the air, soil, and water bodies.

3. Precipitation Precipitation, whether it is rain or snow, is expressed as the depth to which it would cover a horizontal projection of the Earth's surface, if there is no loss by evaporation, run, or infiltration and if any part of the precipitation falling as snow or ice were melted. It is expressed in the units of mm or cm.

Exercise: Answer the following questions.

1. What type of vegetation is dominant in southeast Rajasthan?
2. Which tree species are commonly found in the forests of southeast Rajasthan?
3. What is the main characteristic of the vegetation in the Aravalli hills of southeast Rajasthan?
4. Which of the following vegetation types is found in the riverine areas of southeast Rajasthan?
5. What is the impact of human activities on the vegetation of southeast Rajasthan?
6. Which organisation is responsible for managing and conserving the forests of southeast Rajasthan?
7. What is the significance of the vegetation in southeast Rajasthan regarding biodiversity?
8. How does climate change affect the vegetation of southeast Rajasthan?
9. What is the role of local communities in conserving the vegetation of southeast Rajasthan?
10. What are the main threats to the vegetation of southeast Rajasthan**?**

Test Paper

Chapter-3
Rajasthan: Geography, Climate, and Vegetation

Time Duration: 30 minutes **Maximum Marks: 25**

General Instructions:

1. The paper consists of 12 questions and is divided into four sections, A, B, C and D.
2. All questions are compulsory.
3. Section A comprises MCQS comprising one mark each
4. Section B comprises VSAQS (very short answer questions) carrying three marks each.
5. Section C comprises SAQS (short questions) carrying three marks each.
6. Section D comprises long questions carrying six marks each.

SECTION-A

Exercise 1: Multiple choice questions: **(1*4=4)**

1. What is the climate of Rajasthan?
(a) Tropical desert climate
(b) Hot
(c) Cold
(d) None of the above
2. Western Rajasthan is characterised by
(a) Low rainfall
(b) High rainfall
(c) Very High Rainfall
(d) None of the above
3. Vegetation in Rajasthan
(a) Deciduous
(b) Tropical
(c) Dry Deciduous
(d) All the above
4. State tree of Rajasthan
(a) Prosopis
(b) Banana
(c) Tulsi.
(d) Mango

Answer 1. (a) 2. (a) 3. (a) 4. (a)

SECTION-B

Exercise 2: Very Short answers type questions. **(3*3=9)**

1. What is the typical climate of Rajasthan?
2. List the major vegetation of Rajasthan.
3. What plants are more common in Rajasthan?

SECTION-C

Exercise 3. Short answer type questions. **(3*2=6)**

1. Describe briefly the vegetation of Rajasthan.
2. What type of forests are common in Rajasthan?

SECTION-D

Exercise 4. Long answer type questions. **(6*1=6)**

1. Describe the vegetation and climate of Rajasthan in detail.

Online Resources

	Online Practice Paper	**Online Video Link**
Link	https://forms.gle/ATu6nL4zkyxCiq6x6	https://www.youtube.com/playlist?list=PLzlh_31VWEFrOh3NLkocQ-UhTGLw2Muao
QR Code		

Chapter 4: Methodology for Ethnobotanical Survey

In this chapter, you will learn about

An Ethnobotanical survey was conducted to document the traditional knowledge of plant use among local communities.

The survey aimed to identify and record plant species used for medicinal, food, and other purposes. A survey was conducted in the proposed area of study, and the plants were identified in the Herbarium of the Department of Botany at the University of Rajasthan, Jaipur, using various references. A comparison was also made between the plants, especially herbal plants, in the area identified for study and adjoining areas.

Materials

Field Kit: A Field kit was required during the survey, which consisted of the following articles:

Equipment required for collecting plant specimens during the survey:

a) Plant press
b) Blotting sheets and old newspapers for collection
c) Empty fertiliser bags
d) Polythene bags of various sizes
e) Secateurs
f) Digger
g) Rubber bands
h) Knives
i) Scissors- Big and small
j) String

Equipment required for documentation:

a) Field diaries for record
b) Tape recorder, with external microphone
c) Mobile phone
d) Camera to capture the recordings
e) Photographic film rolls
f) Batteries
g) Extension cord (Electric)
h) Notebooks and writing pads
i) Slip pads for writing notes

Stationary:

a) Paper rim
b) Pen
c) Pencil

d) Sharpener
e) Thread roll
f) Measuring tape
g) Pocket lens
h) Stapler
i) Cello tape roll

Literature:
a) The flora of the area is already available
b) Map and toposheet of the area to be surveyed
c) List of the local plants of the area (derived from the forest department in the area to be studied)
d) List of local names of plants of the area
e) Text on vegetation, flora, people, Ethnography, etc., taken from the
f) District's officers
g) Anthropological publications on the tribe/ community available
h) Ethnobotanical publications on the area/community to be surveyed

Apart from these major equipment and accessories, several miscellaneous items are also to be carried along during the survey, like-
a) Haversacks
b) Hunter shoes
c) Torch
d) Medicines required (for common problems)
e) Candles and match boxes in case of emergency
f) A bag for keeping articles listed above

Secondary Sources
Some facts were gathered from forest officials and personnel of the Tribal Research and Training Institutes in Udaipur, Rajasthan. Additionally, information was collected from researchers, particularly those working in this field at various centres.

Literature as a source
During the surveys, data were also collected from published sources in the fields of agriculture, forestry, and social science, along with botanical and Ethnobotanical information. Some non-botanical literature was also studied:
(i) Anthropological and sociological books and journals.
(ii) Historical aspects of district Dholpur.
(iii) Census publication, etc.

Documentation of Data
In the study region, most activities were documented, and among them, critical data were recorded.

Botanical
Plant parts such as seeds, roots, gums, fruits, leaves, bark, grain, etc. and plant propagules like tubers, culms, corms, bulbils, etc. were also gathered for herbarium specimens.

Exploration
Under this, the floristic survey was conducted in the Dang area of Dholpur, Rajasthan. The floristic survey was conducted during different seasons (frequently three times a year) to ensure accurate identification of the plants. The information regarding the traditional use of medicinal plants was collected through interviews with tribal groups, including women, men, local Vadhaji, and Bhopal residents.

Data Collection
The data were collected using a random sampling method, including interviews, surveys, observations, and participant observations. All information documented from Indigenous people and tribal species was classified into the following data -
(a) Qualitative data, such as applications of plants
(b) Quantitative data, such as qualities of plants in the form of medicine
(c) Descriptive data, which includes information about festivals, rituals, etc., and the role of plants in their life.
(d) Supportive data which involves other information like occupation, vegetation, socio-economic conditions, pastimes and flora.

Survey
The surveys were divided into two types, namely intensive and extensive. Intensive surveys gather information about one or two villages at a time, lasting seven days. Extensive surveys were conducted by establishing a town as a base camp and undertaking extensive work within several villages within its radius. The interviews were conducted with women, men, Gotheeyas, Bhopal residents, local voids, and other indigenous and tribal populations who possess medicinal knowledge. In addition, government physicians, NGOs, college students, forest officials, veterinary doctors, and nurses from villages were also involved in the data collection, as they reside among the tribal communities that possess extensive ethnobotanical knowledge. To gather information, a structured, open-ended questionnaire was prepared and validated by experts.

Observations
In the village, connections were established with or among people, especially with the chief, and subsequently, relationships were formed with other individuals in the area. An expedition to rural areas was also conducted to document the shrines, shelter patterns, and farms, among other features. Investigations of the plant and its parts, as used in various artefacts, were conducted, and related notes were collected.

Interviews
To collect Ethnobotanical information, many tribal people and other persons with Ethnobotanical knowledge were interviewed. Interviews with individuals were conducted in two patterns: one at the individual level and the second at the group level. In the case of personal levels, people were randomly chosen, along with the Gotheeya, to collect information. In group interviews, data collection was conducted with multiple participants. Numerous areas were selected for

interviews. The tribal communities were encouraged to share their knowledge about the medicinal values of plants with local people.

Mode of Interviews

1. A handy checklist of plants likely to occur in the locality, based on personal experience, the forest department's working plans, and regional floras with local names, was used to inquire about their uses.
2. A list of local plants was prepared by enquiring of the tribal or from a prior excursion with a tribal or two accompanying, and enquiries on the enlisted plants were made.
3. An excursion into the forest was made before interviews to collect specimens of all possible species. Then, after showing them the specimens, information was collected.
4. An interview was initiated at random, covering all possible aspects, and plants that figured in the same were noted, which were later collected, taking two or more participants into the forest.
5. The interviews with tribal people were conducted in the forest, and concurrently, ethnomedicinal plant species were collected, along with their applications.

If plants could not be collected, as they were not available, details of their local names, gross morphology, habitat, and other characteristics were recorded during subsequent surveys in different seasons. The interpretations were made after interviewing the local people with the help of structured interviews.

Study design and sampling techniques

Various localities were visited, and data were collected through several techniques, including purposive, snowball, and random sampling.

1. Purposive sampling technique: Screening of individuals with good knowledge and ready to share information with scholars (Cress Well, 2011)

2. Snowball sampling technique: Information gathered from those healers who did not have official permission to share their traditional knowledge about medicinal plants.

3. Random sampling technique: This was the most straightforward technique for data collection. Every member was an internal part of this sampling process.

Interpretation from the interview

Subject: Opening of an interview

Q. 1. What is this? (Plant Identification)

Q. 2. Is this useful?

Q. 3. How? (Give Detailed Use)

Q. 4. Which part is valuable?

Q. 5. Is the Seed useful?

Q. 6. Local Name?

Q. 7. Is it edible?

Q. 8. Any other use?

Q. 9. Any other information about it?

(i) Village

(ii) Date

(iii) Name

(iv) Occupation

A sample performance was prepared in the form of a structured questionnaire to gather information about the plants used by tribal people for curing their health ailments, as well as those of their cattle.

Sample Performa

Personal Details:
a) Name of Herbal practitioner ....
b) Age of Informant ....
c) Tribe practitioner community ....
d) Name of Village ....
e) Occupation of informant ....
f) Year of Experience ....

Q.1 Name of ailments cured.

Q.2 Disease name in the local language.

Q.3 Frequency of disease. Very Frequent/ Rare

Q.4 Name of plants used in treatment.

Q.5 Specific plant part used to cure the disease. Do they use combinations of plants to prepare medicine?
(Specific ratio with details)

Q.6 Plant habitat: wild or cultivated.

Q.7 Frequency of occurrence and growing- (very common, Rare, Endangered).

Q.8 How do you use this medicine?

Q.9 How often do people use it?

Q.10 What do they use it for?

Q.11 What does it do, and how does it affect people?

Q.12 Mode of administration for medicine?

Q.13 How do they name the plant parts in the local language?

Q.14 Plants used exclusively as medicine or used as food and medicine both.

If used as food, then enlist the products: --
-

Q.15 Seasonal availability in the area.

Q.16 What are the major diseases found in the area, and how are they cured?

Q.17 Changes observed during the last 10 years.

Q.18 How did they acquire the Ethnomedicinal knowledge? (Specify the source and mode)

Q. 19 Level of Ethno-medicinal knowledge of the Informant.
(a) Average (b) Good (c) Below average (d) Excellent

Q. 20 Opinion on exposure of Ethnomedicinal knowledge to the young generation.

(Plant Name: --)

Study design and sampling techniques:
Various localities were visited, and data were collected using several techniques, namely, purposive, snowball, and random sampling.
1. Purposive sampling technique: Screening of the individuals with good knowledge and ready to share information with scholars (Cresswell, 2011)
2. Snowball sampling technique: Information gathered from those healers who did not have official permission to share their traditional knowledge about medicinal plants.
3. Random sampling technique: This was the most straightforward technique for data collection. Every member was an internal part of this sampling process.

Summary

An ethnobotanical survey highlights the importance of traditional knowledge and plant use among local communities. A survey provides valuable information for conservation and sustainable development efforts. Further research and collaboration are needed to protect plant diversity and promote sustainable livelihoods.

Glossary

1. Conservation: The protection and preservation of natural resources, including plants and other living organisms.

2. Documentation: The process of recording and preserving traditional knowledge and practices.
3. Folk medicine: Traditional medicine practised by indigenous communities.
4. Indigenous: Originating from a particular region or country.
5. Informant: A person who provides information about traditional knowledge and practices.
6. Local community: A group of people living in a particular area.
7. Methodology: The systematic approach used to collect and analyse data.
8. Participant: A person who takes part in a research study.
9. Phytochemical: A chemical compound produced by a plant.
10. Questionnaire: A set of questions used to collect data.
11. Survey: A research method used to collect data through observations or questionnaires.
12. Traditional knowledge: The knowledge and practices passed down through generations within a community.

Exercise: Answer the following questions.

1. What is the primary objective of an Ethnobotanical survey?
2. Which of the following is a key component of an Ethnobotanical survey?
3. What is the significance of informed consent in an Ethnobotanical survey?
4. How do Ethnobotanical surveys contribute to conservation efforts?
5. What is the role of local communities in an Ethnobotanical survey?
6. How do Ethnobotanical surveys promote cultural preservation?
7. What are the potential challenges of conducting an Ethnobotanical survey?
8. How can the results of an Ethnobotanical survey be used?
9. What is the importance of collaboration with local communities in an Ethnobotanical survey?
10. How can Ethnobotanical surveys contribute to sustainable development?

Test Paper

Chapter 4
Methodology for Ethnobotanical Survey

Time Duration: 30 minutes **Maximum, Marks: 25**

General Instructions:

1. The paper consists of 12 questions and is divided into four sections, A, B, C, and D.
2. All questions are compulsory.
3. Section A comprises MCQS, carrying one mark each.
4. Section B comprises VSAQS (very short answer questions) carrying three marks each.
5. Section C comprises SAQS (short questions) carrying three marks each.
6. Section D comprises long questions carrying six marks each.

SECTION-A

Exercise 1. Multiple choice questions: **(1*4=4)**

1. What is the primary objective of an Ethnobotanical survey?
a) To document traditional knowledge of plant use
b) To identify new plant species
c) To promote conservation of biodiversity
d) To develop new medicines
2. Which of the following is a key component of an Ethnobotanical survey?
a) Collection of plant specimens
b) Interviews with local communities
c) Observation of traditional practices
d) All of the above
3. What is the significance of informed consent in an Ethnobotanical survey?
a) To ensure the participation of local communities
b) To protect intellectual property rights
c) To promote cultural sensitivity
d) All of the above
4. How do Ethnobotanical surveys contribute to conservation efforts?
a) By identifying areas of high conservation value
b) By promoting the sustainable use of plant resources
c) By supporting community-based conservation initiatives
s) All of the above

Answer 1. a) 2. d) 3. d) 4. d)

SECTION-B

Exercise 2. Very Short answer-type questions **(3*3=9)**

1. What is the importance of an Ethnobotanical survey?
2. How is rapport built with tribals during an Ethnobotanical survey?
3. What is the significance of an Ethnobotanical survey in promoting the conservation of plant species in Rajasthan?

SECTION-C

Exercise 3. Short answer type questions- **(3*2=6)**

1. Differentiate between primary and secondary sources of literature.
2. What are the significant roles of the herbarium in Ethnobotanical studies?

SECTION-D

Exercise 4. Long answer type questions. **(6*1=6)**

1. Explain the important steps carried out during an Ethnobotanical survey.

Online Resources

	Online Practice Paper	**Online Video Link**
Link	https://forms.gle/g4GTwkyMq9EpEnC8A	https://www.youtube.com/playlist?list=PLzlh_31VWEFpFTNS7M1B8gShE7sEi7FBV
QR Code		

Chapter 5: Medicinal Plants of Rajasthan

In this chapter, you will learn about:

Applications of Ethnomedicinal plants used by the indigenous people of Rajasthan to cure their health ailments.

1. Definition and meaning of Medico-Ethnobotany
2. Indigenous knowledge about the traditional healthcare practices
3. Use of Plants in Traditional Healthcare Practices.

Rajasthan is a state that harbours a rich biodiversity, where plants are utilised by the tribals and local people of the area for curing various health ailments. The comprehensive table lists the plants used by people, along with their families, the part of the plant used, and the mode of administration for curing health ailments in both people and cattle. A structured questionnaire was prepared to interview the tribal people and gather information about the medicinal importance of plants. The documentation of the plants is given in the following table (Table 1).

Table 1: Ethnomedicinal plants used by the tribal people in curing health ailments

Table 1: Ethnomedicinal plants used by the tribal people in curing health ailments

S. No	Name of Plant	Family	Plant Part	Ailment	Mode of administration
1.	Ammannia baccifera Linn.	Lythraceae	Leaves	Burnt body parts	Fresh leaves are bruised and applied.
2.	Anogeissus pendula Brand.	Combretaceae	Leaves Twigs	Burnt body parts, clean the spots	Decoction is applied.
3.	Amaranthus caudatus Linn.	Amaranthaceae	Leaves	Constipation	The leaves of this plant are eaten as a vegetable.
4.	Ageratum conyzoides Linn.	Asteraceae	Plant	Burns and skin eruptions	Dried powdered plant paste is applied locally.
			Leaf	Conjunctivitis	Leaf juice is given orally with water.
5.	Aegle marmelos (Linn.) Corr.	Rutaceae	Leaf	Dysentery	Ash of a plant leaf, appropriately mixed with curd, and along with black salt, is given orally.
			Root	Protrusion of the uterus	Root pulverised properly with sugar and then taken by women.
			Fruit	Diarrhoea	Immature fruit pulp is chopped into fine pieces,

S. No	Name of Plant	Family	Plant Part	Ailment	Mode of administration
					shade-dried, and then crushed into a powdered form, with one teaspoon used twice a day.
6.	Abelmoschus esculentus (Linn.) Moench	Malvaceae	Fruit	Syphilis, Sexual potential	Decoction taken orally with sugar.
			Root	Leucorrhoea	Decoction taken orally in the morning.
7.	Abrus precatorius Linn.	Fabaceae	Leaves	Heart inflammation	Used orally.
			Roots	Sexual vigour, Leucorrhoea	Oral root decoction is administered twice daily.
			Seeds	Abortion	Orally, seed powder is used by mixing it with water.
			Seeds	Acne, leprosy, ringworm, and pimples	A seed paste, combined with Cannabis and Sesame, is applied wholly to the affected site.
			Seeds	Heating	Overnight, soak the seeds in water collected early in the morning.
8.	Abutilon indicum (Linn.)Sweet.	Malvaceae	Leaves	Diabetes, Syphilis, Dental ailment	Leaves powder mixed with cow's milk is taken orally, Applied Locally, or used as an extract for gargling.
			Leaves	Stone in the urinary bladder	Leaf juice taken orally.
			Whole Plant	Gonorrhoea	Decoction taken orally or applied orally.
			Seeds	Syphilis, Constipation	Paste orally, Seeds powder taken orally.
			Whole plant	Sesexual potential and sperm scarcity	Plant powder, when properly combined with sugar, is used orally.

S. No	Name of Plant	Family	Plant Part	Ailment	Mode of administration
			Bark	Whooping Cough	Decoction orally.
			Unripe fruits/Pods	Leucorrhoea (Red and white)	Taken orally with sugar candy/Sugar.
			Unripe fruits/Pods	Null Sperm, weak sperm, purifying sperm	Unripe and shadow-dried pods are ground with sugar candy and Cardamom and given orally with milk.
10.	Acacia leucophloea (Roxb.)Wild.	Fabaceae	Stem bark	Sexual potency in males and for menstrual complaints in females	The pounded stem bark is soaked overnight in water and taken orally in the morning.
11.	Acacia nilotica (L.) Del.	Fabaceae	Bark, leaves	Wounds	Bark and leaf paste is applied to the affected area.
			Stem bark	Cough, mouth sores	Stem bark is chewed orally with salt or boiled in water and taken orally or gargled.
			Pod	Leucorrhoea	Paste is thoroughly combined with raw sugar and taken on an empty stomach in the early morning.
			Stem bark	Menorrhagia	Stem bark pulverised with sugar and leaves of Dalbergia sissoo, and given early morning on an empty stomach. It can be preserved in an earthen pot.
12.	Acalypha indica Linn.	Euphorbiaceae	Whole plant	Brain weakness	Extract given orally.
			Fresh or dry plant	Gastro-intestinal irritation	Extract given orally.

S. No	Name of Plant	Family	Plant Part	Ailment	Mode of administration
			Roots	Constipation	Powder with cow milk
			Whole plant	Cold and cough	Decoction mixed with Jaggery.
			Roots	Scabies, cutaneous eruptions	Paste is applied locally.
			Whole plant	Pneumonia, asthma	Decoction taken orally thrice a day.
13.	Achyranthes aspera Linn.	Amaranthaceae	Roots	For easy delivery (childbirth)	Put under the head/legs of women.
			Roots	Chronic and old-time headache	Root paste was administered orally early in the morning for four consecutive days, accompanied by jaggery.
			Seeds	Cough	Seed powder is thoroughly combined with black sugar and taken orally three times a day.
			Inflorescence	Scorpion sting	Paste is applied locally.
			Whole plant	Cough and cold	Decoction given orally.
			Whole plant	Check for excessive bleeding during delivery	Extract taken orally.
			Roots	Bleeding from cuts	Paste of roots applied locally.
			Roots	Piles and fissures	Paste locally
			Roots	Pneumonia	Root decoction orally taken for three days with jaggery.
			Root	Gangrene,	Orally used root

S. No	Name of Plant	Family	Plant Part	Ailment	Mode of administration
				Typhoid, Rheumatism	decoction.
			Root (Fresh)	Mild fever, especially bone fever	Root decoction is taken orally daily (two times a day).
			Leaves	Rickets	The leaves are pasted correctly, combined with jaggery, and this paste is applied entirely to the head.
			Roots	Cramps	Pounded roots taken twice a day.
14.	Aerva lantana (Linn.) Juss, ex Schult.	Amaranthaceae	Inflorescence	Snake bite	Decoction topically.
			Whole plant	Chest pain	Decoction given orally.
15.	Allium cepa Linn.	Amaryllidaceae	Bulb	Sexual potential	Taken orally with ghee.
			Bulb	Scorpion sting	Paste topically.
			Bulb	Protrusion of the uterus	Boil with ghee and give orally for five days.
			Juice	Urinary Tract Infection (U.T.I.) and burning	Juice with water given orally for two days.
			Bulb	Unconsciousness	Crushed and inhaled.
			Bulb	Irregular menstruation	Eaten orally excessively.
16.	Allium sativum Linn.	Amaryllidaceae	Bulb	Sexual potential	Orally given with ghee.
			Bulb	Scorpion sting	PaStep topically.
			Bulb	Epilepsy	Piece eaten daily with milk.
			Bulb	Ringworm	Paste is applied externally for two days.
			Bulb	Cough and Cold	The roasted bulbs are eaten directly.
			Bulb	Fever and a cold cough	Crushed bulbs are given twice a day for seven

S. No	Name of Plant	Family	Plant Part	Ailment	Mode of administration
					days.
			Bulb	Bulbs	Bulbs are given orally for three days.
			Bulb	Cough and Asthma	Two bulbs were eaten directly.
			Bulb	Eczema	Roasted bulb in dalda ghee, then mixed with lemon juice and applied topically daily.
17.	Adhatoda zeylanica Medic.	Acanthaceae	Leaves	Asthma, cough	Decoction of leaves is given orally.
18.	Adina cordifolia (Willd. Ex Roxb.) Hook. F. ex Brandis	Rubiaceae	Bark	Stomachache	Fresh bark is ground with brown sugar.
			Stem bark	Rheumatism	Fresh stem bark juice.
19.	Actiniopteris radiata (J.Koenig ex Sw.) Link.	Pteridaceae	Whole plant	Nocturnal emission	The entire plant was soaked in water for twelve hours and then crushed gently with milk.
20.	Aloe barbadensis Mill.	Asphodelaceae	Leaves	Sexual vitality	Taken orally.
			Leaves	Asthma	Pulp mixed with ghee and taken orally.
			Leaves	Alopecia	Pulp mixed with paste of Anacardium leaves is applied locally.
			Leaves	Weakness	Pulp mixed with flour, sugar and ghee, taken orally.
			Stem	Burns	Extract locally.
			Leaves	Rheumatism, Arthritis	Pulp mixed with flour makes chapatti, which is eaten orally, or pulp mixed with milk is taken orally.

S. No	Name of Plant	Family	Plant Part	Ailment	Mode of administration
			Leaves	Joint pain	Aloe vera can be taken orally with milk.
21.	Asparagus racemosus Willd.	Liliaceae	Fruits	Lactation inducer	Orally taken pounded fruits up to seven days.
			Leaves	Paralysis	Taken orally.
			Stem	Chronic Dysentery	Stem extract is given orally.
			Roots	Sexual potential	Root powder is mixed with milk and administered orally.
			Roots	Galactogougae	Orally taken root juice
			Roots	Leucorrhoea	Root powder entirely mixed with ghee and given orally.
			Roots	Diabetes, Bronchitis	Decoction taken orally.
			Roots	Weakness	Powder mixed with Withania powder and given orally.
			Roots	Epistaxis	Root juice is injected into the nose dropwise.
			Roots	Diarrhoea and dysentery	Powder taken with water orally.
			Pods	Sprain, Bone ache	Two grams of pod powder mixed with 250 mL of milk is given orally in the early morning.
			Roots	Lactation in women and animals	Fresh root eaten orally with black sugar and milk
			Pods	Sexual weakness	Pod powder combined with Sphaeranthus indicus and ginger, and given with wheat flour and ghee, is consumed daily as part of breakfast.

S. No	Name of Plant	Family	Plant Part	Ailment	Mode of administration
			Roots	Gout, Rheumatism	Taken orally as tea.
22.	Asphodelus tenuifolius Cav.	Asphodelaceae	Seeds	Ulcer and inflammation	Paste applied locally.
23.	Adhatoda vasica Nees.	Acanthaceae	Leaves	Cough and cold	Decoction of the leaves.
24.	Azadirachta indica A, Juss.	Meliaceae	Leaves	Heel pain	50 gm leaves tied on the heel, warmed bricks put on the heel and pour the mattha on the brick for three days during the process.
			Flower	Irregular menses	Paste is applied locally.
			Leaves	Snake bite, Boils	Paste is applied locally.
			Leaves	Dental ailments	Extract to gargle.
			Leaves	Malaria or intermittent fever	Infusion.
			Bark	Piles	Paste with honey taken orally.
			Bark	Diarrhoea	Decoction with honey taken orally.
			Bark	Abdominal pain	Decoction with honey.
			Leaves	Skin disease dermatitis	A paste of leaves and black sugar is taken orally for five days.
			Neemjhar or flower inflorescence	Fever vaccine	Make a Rayata with Glolua and Neem jhar and drink directly.
			Leaves	Acne/pimples	Leaves roasted half and filtered and given orally.
			Endosperm of seed	Blooded piles	Endosperm paste mixed with Solanum nigrum and taken orally.

S. No	Name of Plant	Family	Plant Part	Ailment	Mode of administration
			Leaves	Skin and Pimples or Dermatitis	Paste applied locally (externally) on pimples daily.
			Leaves	Germicide	Given orally, leaves crushed in warm water and wash the vulva of delivered cattle.
			Bark	Irregular menses Scabies and wounds	Decoction with honey given orally. Ash mixed with coconut oil and applied locally.
			Twigs	Dental ailments (pyrrohoea)	Used as toothbrush.
			Seeds	Birth control	Oil, locally applied.
			Fruits	Blood impurity	Eaten orally.
			Barks	Ghamoria	Paste with water applied externally.
			Roots Barks	Anti acne valgeris	Paste with Santalum album and Curcuma longa applied locally.
			Roots	Gastric pain due to gastritis	Decoction with jaggery and water given orally.
			Fruit/seeds	Leucorrhoea (white discharge)	Endosperm mixed with deshi boora and eaten directly for three days, no bath during therapy.
			Leaves	Crack heels	Extract of leaves taken orally with deshi boora daily in morning for fitness.
			Leaves	Eye flue, Conjunctivitis, Eye pain, Eye burning	Extract of leaves dropped in open eyes in this disease.
			Leaves	Weak eye sight	Make a kajal by dried and burned leaves ash mixed with mustared oil and

S. No	Name of Plant	Family	Plant Part	Ailment	Mode of administration
					apply locally in open eye lids.
			Fresh Leaves	Insects/worms in intestine	Fresh leaves chewed directly daily in morning.
			Leaves and seed oil	Antifungal, antimicrobes infection	Seeds oil can be used as spray for micro-infection
			Leaves	Sprain	Tied directly with ghee on affected area
			Twigs with leaves	Vaginal infection in animal after delivery	Leaves wash by warm water.
			Leaves	Malaria	Paste of leaves mixed with Alum and make tablet, taken orally before fever and after fever daily for three days.
			Leaflets	Cholera	Decoction of leaflet combined with Ellachi piper and ash of Cocos nucifera and water, taken orally for fitness.
			Seed oil	Scabies	Oil applied on infected site of body externally.
25.	Alstonia scholaris	Apocynaceae	Stem, bark	Pain during delivery	Decoction with country liquor.
26.	Atylosia scarabaeoides	Leguminosae	Root	Rheumatism	Paste.
27.	Actiniopteris radiata (J.koenig ex Sw.) Link.	Pteridaceae	Whole plant	Nocturnal emission, tonic	Entire plant soaked for tweleve hours in water and taken with milk.
28.	Antigonon	Polygonaceae	Leaves	Blisters	Leaves are crushed with

S. No	Name of Plant	Family	Plant Part	Ailment	Mode of administration
	leptopus Hook. & Arn.				sugar and applied twice or thrice for a day.
29.	Argemone Mexicana Linn.	Papaveraceae	Leaves	Malaria fever, ulcer and skin	Leaves decoction.
			Stem, leaves	Gastralgia, asthma and cough	The leaves and stem decoction are used three times a day.
30.	Aristolochia indica Linn.	Aristolochiaceae	Leaves	Rheumatic pain, sting of scorpion and insect bite	Paste of fresh leaves is applied.
			Root	Collapse and snake bite	Root paste is taken orally with water.
31.	Annona squamosa Linn.	Annonaceae	Seeds, leaves	Abortion	Seed powder combined with Plumbago zeylanica leaves.
32.	Albizia lebbeck (Linn.) Benth.	Leguminosae	Stem bark	Ulcer, insect bite and scorpion sting	Leaves and stem bark mixed and formed paste and then applied on the affected sites.
33.	Alangium salvifolium (L.f) Benth.	Cornaceae	Stem bark	Uterus related ailments	The stem bark of the plant is crushed with Bombax flowers, and this mixture is taken twice a day orally.
34.	Bacopa monnieri (Linn.) Wettest.	Plantaginaceae	Flowers and leaves	General disability	Leaves and flower decoction combined with milk and honey and then takenorally.
35.	Bauhinia purpurea Linn.	Fabaceae	Flower	Constipation	Decoction of flower buds used orally.
36.	Balanites aegyptiaca Linn.	Zygophyllaceae	Fruits	Birth control	Powder taken orally.
			Fruits	Whooping cough	Pulp taken orally.
			Fruits	Burns, ulcer and eruptions	Pulp applied locally.
			Roots	Dysentery	Extract taken orally.

S. No	Name of Plant	Family	Plant Part	Ailment	Mode of administration
			Roots	Internal haemorrhages	Paste rubbed locally.
			Roots	Guineaworm	Paste given orally.
			Seeds	Dysentery	Seeds dried and burned . The ash is taken orally with honey.
			Seeds	Gastritis	Powder given orally.
			Roots	Haemorrhages	Pounded roots dipped overnight in water and orally taken early morning.
			Whole plant and seed	Protrusion of the uterus and inflammation	Whole plant with fruits, dipped in water 5-10 times continuously, and administered orally.
37.	Boerhavia diffusa Linn.	Nyctaginacea e	Leaves	Kidney stone	Used in the form of a vegetable.
			Roots	Jaundice	Root paste is orally used.
			Roots	Kidney stone	Root powder is given orally with water.
			Roots	Fever	Root extract is given orally with honey.
			Whole Plant	Sexual vigour	Decoction of the entire plant is formed by using other plants such as Solanum, Eclipta and given orally.
			Whole Plant	Loss of appetite	Prepared decoction of this whole plant with chandan and taken orally.
			Roots	Eye itching	A paste is formed by mixing roots with milk and poured into the eyes dropwise.
			Roots	Water/tear secretion in the eye always	Root mixed with water and then used.

S. No	Name of Plant	Family	Plant Part	Ailment	Mode of administration
			Roots	Flower of the eye (white spot in the eye)	Dragged in ghee applied locally.
			Roots	Eye ailment (Timir)	Root powder mixed with sesame and used.
			Roots	Night blindness	Powder mixed with "Kanji" and used.
			Roots	Teething (the first teeth of newborns) problem in infants	Root infusion applied to the neck with a blue cloth(on Saturdays)
			Roots	Jaundice	Inhale the smell of fresh roots up to eight days.
			Roots	Jaundice	Eaten orally with Sugarcane juice.
			Roots	Jaundice	Root infusion applied to the neck by using a thread (red colored).
			Roots/ leaves	Jaundice	Eaten with Cicer arietinum orally.
			Roots	Moderate/ mild fever	Root powder with water i given orally twice a day.
			Plants	Urine problem	Aerial parts are given orally for five days.
38.	Boswellia serrata Roxb. Ex Cocls.	Burseraceae	Leaves	Scorpion sting	Paste topically.
			Bark, gums and resins	Abscesses	Locally applied.
			Leaves and bark	Body pain	Decoction taken orally.
39.	Brassica campestris Linn.	Brassicaceae	Seed oil	Eye sight weakness	Oil dropped in eye at bed time daily.
			Seeds	Stomachache in a child	50 gm seeds mixed with hing, make a paste and apply locally on the navel.

S. No	Name of Plant	Family	Plant Part	Ailment	Mode of administration
40.	Brassica juncea Linn.	Brassicaceae	Seeds oil	Ghamoria	Mix with cold water and apply to the back or affected body area for regular massage.
41.	Buchanania latifolia Roxb.	Anacardiaceae	Stem bark	Infants	Powder of the stem bark with Syzygium cumini.
42.	Barleria cuspidata Linn.	Acanthaceae	Leaf	Rubbed on painful areas, the muscular pain	The leaf juice of this plant is mixed with cooking oil.
43.	Bombax ceiba Linn.	Malvaceae	Flower	Acidity and urinary problems	Flower infusion with sugar taken orally.
			Root, fruit	Calm body heat	Powder of root mixed with Capparis sepiaria, Chlorophytum and P. murex fruits used as a tonic with water.
44.	Bauhinia variegata Linn.	Fabaceae	Leaf	Flowing ear	Powder is externally applied with mustard oil.
45.	Barleria prionitis Linn.	Acanthaceae	Leaves	Paralysis	Leaves (fresh) boiled and prepared a paste and tied with a cotton bandage, and the filtrate is poured on the affected part twice a day for a month.
			Root	Rheumatic fever	Paste of roots mixed with goat milk.
46.	Butea monosperma (Lam.) Taub.	Fabaceae	Fruits	Sprains and fractures	Roasted fruits are tied locally.
			Gum	Birth control	Gum used orally with water up to seven days.
			Gum	Bronchitis	Roasted and used orally.
			Gum	Chronic dysentery	10 gm gum with Mangifera indica and moringa gum, and given orally.
			Flower	Ringworm	Paste of flower is

S. No	Name of Plant	Family	Plant Part	Ailment	Mode of administration
					appliedon the affected sites.
			Flower	Leprosy	Externally applied flower paste.
			Flower	Blood impurity	Flower powder is orally given by mixing with water.
			Flower	Weaken fever	Powder combined properly with sugar and milkand used twice a day.
			Leaves	Guinea worm	Roasted in hot water and tied locally.
			Seeds	Intestinal helminthes	One or two seeds orally given.
			Stem bark	Protrusion of uterus	Orally used decoction of stem bark.
			Stem bark	Dental ailment	Used stem bark decoction for gargle.
			Stem bark	Antidote to itching	Scrubbed over effective area.
			Leaves	Abscesses, Phoda	Leaves (roasted) tied properly on affected organ.
			Seeds	Birth control	Three/five seeds orally given to women in menstruation time up to five days.
47.	Bauhinia variegata Linn.	Fabaceae	Root	Mucilage and gums	Decoction of root.
48.	Cordia myxa Linn.	Boraginaceae	Pulp	Rheumatic pain	Pulp applied.
49.	Calotropis gigantea (Linn.) R. Br.	Apocynaceae	Root bark and leaves	Snake bite	Paste with ox urine, topically.
			Roots	Cold and cough	Extract orally.
			Leaves	Abscess	Crushed leaves locally.

S. No	Name of Plant	Family	Plant Part	Ailment	Mode of administration
			Leaves and roots	Fever, leprosy, chronic rhumatism and syphilitic ulceration	Extract orally.
			Latex	Guinea worm	Locally applied.
			Latex	Diarrhoea and dysentery	Orally given twice a day.
			Leaves	Swellings	Warmed leaves are applied thrice in a day.
			Whole plant	Jahar aana	Paste made with black sugar and given orally for three days.
			Leaves	Scabies	Paste applied locally.
			Whole plant	Eczema and skin disease	Paste applied locally.
			Whole plant	Gastric ailments	Decoction taken orally.
50.	Calotropis procera acut. Non (Ait) Ait. f.	Apocynaceae	Leaves	Migraine	Make the tablets with black sugar and take orally twice a day.
			Twigs	Malaria	Two twigs paste mixed with mishree, ilayachi given with buffalo milk orally.
			Roots	Cough and cold	One gram paste of roots mixed with honey and given with rice.
			Latex	Arthritis	Make the vegetable of boiled fresh leaves and eaten with chapatti.
			Latex	To stimulate the vomiting	Two drop orally with batasha given orally.
			Latex	Paranochia on finger tips	Latex mixed with snake skin cover and sindoor, applied locally.
			Latex	Eczema	Latex with mobile oil

S. No	Name of Plant	Family	Plant Part	Ailment	Mode of administration
					applied locally on affected place.
			Latex	Scorpion sting/ Guinea worm	Latex mixed with Stinging nettle given orally and applied locally.
			Leaves	Thorn	Latex applied on thorn locally.
			Leaves	Abdominal pain	Yellow leaves (warmed); are tied on abdomen, locally poultice over abdomen.
			Latex	Sprain	Roasted half leaves tied on sprain.
			Latex	Eczema	Cotton soaked in latex is burnt and later ash is applied locally with mustard oil on affected areas.
			Roots	Scorpion bite	Make the paste of fresh roots and mixed with common salt and caustic soda applied locally on stinged part of body.
			Roots	Asthma	Root stock roasted and mixed with bajara flour and baked as bread eaten orally.
			Leaves	Snake bite	Sixteen leaves (powdery soft) should be taken from different sixteen plants separately (one from each) which is called "Solahari" in language of Hakims and make a paste with 25 gm 'Gur' and given orally to snake biten patient thirty

S. No	Name of Plant	Family	Plant Part	Ailment	Mode of administration
					times and at 15-30 minutes interval but the patient should not be allowed to sleep and to drink water.
			Roots	Vomiting due to snake biting	Decoction of roots given orally two - three times.
			Latex	Paronychia	Paste with black snake skin and sindoor and apply on tips locally.
			Pistil of flower	Mouth cancer	One pistle taken orally with water for fifteen days.
51.	Capparis decidua (Forsk.) Edgew.	Capparaceae	Leaves	Boils and swelling	Paste locally.
			Fruits	Cardiac trouble	Taken orally.
			Bark	Cough and asthma	Decoction orally.
			Roots and roots bark	Rheumatism, intermittent fever and dropsy	Decoction taken orally.
			Twigs	Toothache	As tooth brush.
			Fresh, new borne leaves	Ringworm	Paste applied locally in affected area.
			Leaves	Toothache	Juice/extract dropped in same side of the ear as that of tooth ache side.
			Buds	Sprain	Paste mixed with cow urine, externally.
			Latex	Embedded thorn in body	Latex applied locally.
			Fruit	Indigestion	Used as energy sources
52.	Centella asiatica Linn.	Apiaceae	Leaf	Improving memory	Leaf paste and brown sugar.
53.	Chrozopho	Euphorbiacea	Leaves	Sunburn, facial	Crushed leaves are

S. No	Name of Plant	Family	Plant Part	Ailment	Mode of administration
	ra rottleri (Geis.) A, Juss, ex Spreng.	e		swelling, sun stroke	applied to the face.
54.	Cissampelos pareira auct. Non. Linn	Menispermaceae	Root	Diarrhoea, bronchitis, liver congestion, dyspepsia, pneumonia	Root extract is taken orally.
55.	Cissus quadrangularis Linn.	Vitaceae	Whole plant	Fractured bone, asthma, epistaxis	Milk and whole plant juice is given orally.
			Stem	Fractured bone	Stem is blend into paste and given with cow milk three times in a day.
56.	Citrus aurantifolia Linn.	Rutaceae	Juice	Asthma	Twenty one cowries mixed with juice of 200 ml of lemons and kept till dissolution, filtered and one spoon is used daily.
57.	Citrus medica Linn.	Rutaceae	Fruit juice	Ringworm, scabies, eczema	Fruit juice and water is mixed in1:1 ratio stirred till it becomes like a jelly and applied.
58.	Cleome gynandra Linn.	Cleomaceae	Plant	Skin disease	Plant is crushed in to paste, squeezed and extract is applied.
			Leaves	Colic, intestinal worms, diarrhoea, dyspepsia	Fresh leaves extract taken orally.
			Leaves	Ear complaints	Bruished of fresh leaves decoction is combined with sesame oil and poured in ears.
59.	Cleome viscosa Linn.	Cleomaceae	Leaves	Ringworm, scabies	Fresh paste of stem and leaves is applied locally.
			Root	Colic, flatulence and dyspepsia	Root and candy extract syrup is prepared and

S. No	Name of Plant	Family	Plant Part	Ailment	Mode of administration
					taken orally
60.	Cardiospermum halicacabum Linn.	Sapindaceae	Seeds	Rheumatism	Powder with water or cow milk orally.
			Plants	Tumors	Decoction taken orally.
			Fruits	Internally heating in body	Mixed with cow milk and taken orally.
			Fruits	Leucorrhoea	Taken orally daily with fresh cow milk.
			Fruits	Piles	One fruit mixed with Areca catechu, mishri and Honey and taken orally in morning.
			Fruits	Riddances the smoking habit	Chewing seeds besides Bidi, cigarette.
			Seeds	Mouth freshner	Chewed orally.
61.	Carica papaya Linn.	Caricaceae	Roots	Stone in kidney (Calculus)	5-6 gm roots paste prepared and mixed with water and taken orally in morning for three weeks, stone is dissolved and passed through urinal passage.
62.	Cassia fistula Linn.	Fabaceae	Flower	Rat bite	Juice with cow milk taken orally.
			Leaves	Facial paralysis and Rheumatism	Decoction orally.
			Pods	Anti-helminthes	Pulp taken orally.
			Fruit pulp	Constipation	Taken orally.
			Stem bark	Scabies	Paste locally.
			Roots	Cold and fever	Decoction orally.
			Seeds	Diabetes and snake bite	Taken orally.
			Pulp	Indigestion	Pulp taken orally.
			Stem bark	Diarrhoea	Blend bark stem soaked in water overnight and one given orally in the morning for five days.

S. No	Name of Plant	Family	Plant Part	Ailment	Mode of administration
			Leaves	Pneumonia	Mixed with red chilli and given orally.
63.	Cassia occidentals Linn.	Fabaceae	Leaves	Constipation	Powder taken orally with water.
			Leaves	Ringworm	Extract locally.
			Seeds	Ringworm	Powder locally.
			Leaf	Conjunctivitis	Extract poured or dropped in eye two - three times daily.
			Leaves	Bone fracture	Crushed with egg albumin and sheep milk tied locally on fractured part.
			Leaves	Insect sting	Fresh leaves taken and crushed with extract of this plant tied locally.
			Roots	Scorpion sting	Root taken and chewed fresh by Vaidhya and then ear of patient is blown by him in case of scorpion sting.
			Seeds	Ring worm	Paste of seed applied locally.
			Seeds	Leprosy	Seed of this plant mixed with seed of radish and applied locally.
64.	Cassia pumila Lam.	Caesalpiniacea e	Whole plants	Galactagogue	Plant parts given orally in morning for fifteen days.
			Stem and Whole plant	Anti-rabies, hydrophobia (antidotes)	Juice of plant mixed in ghee and given orally.
			Seeds	Galactagogue	Crushed and soaked in water overnight and given orally in the morning for fifteen days.
65.	Cassia tora Linn.	Fabaceae	Leaves	Constipation	Juice taken orally.
			Leaves	Ringworm	Extract locally.

S. No	Name of Plant	Family	Plant Part	Ailment	Mode of administration
			Seeds	Skin diseases	Powder taken orally.
			Roots	Snake bite	Paste locally.
			Seeds	Excema	Cow urine paste is used locally.
			Seeds	Ringworm	Paste with water used locally.
			Leaves	Night blindness	Used as vegetable or eaten directly for ten days.
			Leaves	Antifungal and Bacterial disease	Extract applied locally.
66.	Centella asiatica (Linn.) Urban.	Apiaceae	Leaves	Headache	Three gm powder with tea leaf taken orally.
			Leaves	Cholera	Decoction orally thrice a day.
			Plant	Malaria intermittent fever, Nervous disorders Nervous disability insomnia	Decoction taken orally.
			Plants	Hair falling	Decoction or paste mixed with Emblica species applied for washing hair.
67.	Cicer arietinum Linn.	Fabaceae	Seeds	Leucorrhoea and menorrhoea or sudden bleeding from vagina	500 gm cicer 250 gm kathiya supari (beetal nut mixed with 250 gm mishree and taken orally for ten days.
			Seeds	Diabetes	Fried or soaked seeds eaten regularly in diet.
68.	Citrullus colocynthis (L.) Schard.	Cucurbitaceae	Fruits	Gastritis	Powder taken orally.
			Fruits	Abdominal pain	Orally with common salt.
			Fruits	Gastritis in animal	Decoction orally.
			Fruits	Constipation	Combined with Triphala

S. No	Name of Plant	Family	Plant Part	Ailment	Mode of administration
					and give orally with water or milk.
			Roots	Rheumatism, Jaundice, Urinary troubles	Extract orally.
			Roots	Gastritits	Terminalia chebula (Haritki) mixed with black salt and ginger and given orally.
			Seeds	Bleeding, Abortifacient	Orally used.
			Roots	Constipation	Dried powder taken orally with milk or water.
			Fruits	Constipation	Use as sauces.
			Fruits	Eye sight weakness	Eaten without seeds orally.
			Ripen fruits	Diabetes	Distilledand fermented extract of fruits taken orally before meal.
			Fruits	Vomiting	One gm jeera, One gm choti ilayachi mixed with juice of Citrus species and given orally at regular interval of 2-2 hours vomiting will be stopped.
			Fruits	For clean teeth	Brush with common salt and juice daily.
			Fruits	Morning sickness	25 gm jaggery soaked over night.In morning 25 gm juice of Citrus mixed in jaggery water and taken orally daily for two-three weeks.
			Fruits	Acne	Juice rubbed on face locally.
			Fruits	Wrinkles	Juice mixed with honey and applied locally on

S. No	Name of Plant	Family	Plant Part	Ailment	Mode of administration
					face.
			Fruits	Loss of appetite	Juice given orally with black salt.
			Fruit	Cholera	Juice mixed with mishri given before meal.
			Fruit	Obesity	Juice mixed with honey and given orally with warm water before meal.
			Fruit	Splenomegaly	Juice mixed with onion juice and given orally for fourteen days.
			Seeds	Scorpion sting	Nine gm of seed kernel mix with saindha namak and taken orally.
69.	Cleome viscosa Linn.	Cleomaceae	Leaves	Burn and cuts	Extract applied locally.
			Leaves	Boils	Paste applied locally.
			Leaves	Ear ailments	Extract applied locally.
70.	Cleome gynandra Linn.	Cleomaceae	Leaves	Burn and cuts	Extract applied locally.
			Leaves	Armpit tumour	Roasted half leaves tied locally.
71.	Clerodendrum phlomidis Linn.	Lamiaceae	Leaves	Syphilis	Juice applied locally.
			Leaves	Anti-helminthic	Juice taken orally.
			Leaves	For curing pinworm	Paste applied locally.
			Roots	Gonorrhoea	Decoction taken orally.
72.	Clitoria ternatea	Fabaceae	Root	Dog bite	Powder with cow milk taken orally.
			Seeds	Syphilis	Paste applied locally.
			Seeds and Leaves	Mental weakness	Decoction taken orally.
			Whole plant	Kak Bandhya (infertility appear after 1st delivery)	Two gm Nagkesar with ten gm paste and given orally with honey for ten days in morning.
73.	Cocculus hirsutus Linn.	Menispermaceae	Leaves	Sexual vigor	Paste taken orally with milk.
			Leaves	Cuts, sores and	Juice applied locally.

S. No	Name of Plant	Family	Plant Part	Ailment	Mode of administration
				wounds	
			Leaves	Weakness	Extract taken orally with sugar.
			Leaves	Eye inflammation	Juice dropped in open eye.
			Roots	Insect bite	Decoction taken orally.
			Stem	Stomach pain	Powder taken orally.
74.	Commelina benghalensis Linn.	Commelinaceae	Leaves	Eye flu	Extract dropped in open eye twice a day.
75.	Commelina forskaolii	Commelinaceae	Stem	Snake bite by king cobra	Paste given orally.
			Roots	Bitten by stripped snake	Paste given orally.
			Leaves	Bitten by yellow color species of snake	Paste given orally.
			Seeds	Skin infection as scabies, scratching lesion on skin due to poison of snake	Paste with cow ghee given orally.
76.	Commiphora wightii (Arn.) Bhandari	Burseraceae	Bark	Bronchitis	Bark powder taken orally with milk.
			Gum	Asthma	Powder taken orally in milk.
			Gum	Pyorrhoea	Gum lotion in water is used for gargles.
			Oleo gum resin.	Piles	Applied locally.
			Twigs	Pyorrhoea	Tooth brush for brushing.
77.	Corchorus aestuans Linn.	Malvaceae	Whole plant	Sexual vigor	Powder taken with fresh goat milk.
			Seed	Mental weakness	Pounded seeds taken in milk.
			Whole plant	Scanty semen	Shankhpushpi, zinger (Sonth) (dried) and Gourak mundi with milk

S. No	Name of Plant	Family	Plant Part	Ailment	Mode of administration
					taken orally.
78.	Corchorus olitorius Linn.	Malvaceae	Seeds	Brain weakness	Powder taken orally with milk.
			Whole plant	Sexual vigor	Powder (dried) mixed with Gorakh mundi and taken orally with milk.
			Whole plant	Internal inflammation in body	Taken orally with sugar and flour daily in fasting.
79.	Corchorus depressus (Linn.) Stocks.	Malvaceae	Roots	Excessive bleeding in menstrual cycle	Paste taken orally with milk.
			Roots	Brain weakness	Decoction orally with milk.
			Seeds	Brain weakness	Pounded seeds taken orally with milk.
			Whole Plants	Protrusion of uterus / internal heat	One gm shadow dried whole plant powder made is given orally with processed curd and mishree (processed sugar) for seven days.
80.	Cordia dichotoma Forst. F	Boraginaceae	Fruits	Weakness	Crushed in water and taken orally.
			Seeds	Ringworm	Paste locally
			Leaves	Burns	Ash of leaves with Sesame oilapplied locally.
81.	Curcuma longa Linn.	Zingiberaceae	Rhizome	Bronchitis upper respiratory tract infection (URI)	Haldi powder mixed with Common salt – 50 gm, Meetha soda – 50 gm, Fitkari – 50 gm,and the powder (one spoon) given orally twice a day.
			Rhizome	Face pack	Powder half spoon mixed with maida (Flour of wheat) honey and juice of Citrus and mixed with

S. No	Name of Plant	Family	Plant Part	Ailment	Mode of administration
					milk and applied locally at bed time daily.
			Fruits	Wrinkled face	Paste of fruits applied locally on face with gulabjal and washed after half an hour with warmwater.
			Leaves	Bald head/ Alopecia	Fresh leaves paste mixed with Aloevera pulp and take bath daily.
			Rhizome	Whooping cough	Powder taken with milk orally.
			Rhizome	Fractured bones	Mixed with chuna and applied locally.
			Rhizome	Cough/ cold	Powder mixed with milk, common salt and given orally.
			Rhizome	Sprain	Brinjal fruits cut into small piece and roasted half in old ghee and turmeric and then tied on the poultice on affected part in sprain for three days.
			Rhizome	Eye inflammation	Powder or paste applied in eye with ghee and cotton plugged.
			Rhizome	Asthma	Powder taken orally with milk.
			Rhizome	Infertility	50 gm powder taken orally with cold water for immunity in menstrual period from first to third day.
			Rhizome	Asthma	Fresh unripe rhizome mixed with chuna and used orally with water for fifteen days.
			Rhizome	Sty (lump) in	Powder of turmeric

S. No	Name of Plant	Family	Plant Part	Ailment	Mode of administration
				eye	combined with Piper longum and applied locally.
			Rhizome	Swelling due to urine of insects	Paste applied locally.
			Rhizome	Cough and bronchitis	Powder of turmeric half spoon mixed with half spoon common salt and taken orally at bed time with warm milk.
			Rhizome	Knee pain	Powder of turmeric mixed with lime and honey and paste applied locally.
			Rhizome	Throat, block due to cold	50 gm turmeric powder, 50 gm meetha soda, fitkari mixed and used for gargle.
82.	Cucumis callosus (Rott.) Cogn.	Cucurbitaceae	Fruits	Diabetes	Juice mixed with amla and gwarapatha juice and given orally.
			Fruits	Gastritis	Pulp of this fruit mixed with black salt, common salt and sugar and given orally.
83.	Cucumis melo Linn. var utilissimus Duthie and Fuller	Cucurbitaceae	Roots	Birth control	Root powder with jaggery given orally.
			Roots	Scorpion sting	Paste applied topically.
84.	Cucurbita moschata (Duch. Ex Lam.)	Cucurbitaceae	Stalk of fruit	Scorpion sting	Sap applied locally.
85.	Cuminum cyminum Linn.	Apiaceae	Seeds	Excessive bleeding in menstrual cycle	Decoction taken orally with sugar.

S. No	Name of Plant	Family	Plant Part	Ailment	Mode of administration
			Seeds	Indigestion	Roasted half seed's powder mixed with mattha taken orally.
			Seeds	U.T.I./ burning	Powder with mishree taken orally.
			Seeds	Clearing of uterus	Paste with water taken orally in morning.
86.	Curcuma aromatica	Zingiberaceae	Rhizome	Pyorrhoea (plaque)	Rhizome powder mixed with Lahori namak and mustard oil used as dental paste for brushing.
87.	Cuscuta reflexa Roxb.	Convolvulaceae	Whole plant	Chest pain	Rubbed and applied locally.
			Stem	Jaundice	Decoction taken orally.
			Whole plant	Sprain	Roasted half plants applied locally.
88.	Cynodon dactylon (Linn.) Pers.	Poaceae	Whole plant	Wounds, Piles	Paste applied locally.
			Whole plant	To fix pregnancy or bleeding during pregnancy	Ground with sugar and taken orally for seven days.
			Whole plant	Eye diseases, vomiting (bleeding)	Juice applied locally.
			Whole plant	Tuberculosis	Juice taken orally in morning for one month.
			Whole plant or leaves	Boils	Apply the paste locally.
			Juice	Epistaxis	Juice dropped in nasal pore.
89.	Curculigo orchioiides Gaertn.	Hypoxidaceae	Roots	Loss of appetite, indigestion, constipation, dysentery, gastritis abdominal pain	Seed of jeera mixed with kishmish, anardana, black salt, and mishree grounded with juice of citrus, make tablet 2-2-tablets taken orallydaily

S. No	Name of Plant	Family	Plant Part	Ailment	Mode of administration
					after meals.
			Seeds	U.T.I.	Seed powder given orally with sugar or honey.
90.	Cannabis sativa Linn.	Cannabaceae	Leaves	Stomachache	In coconut milk bruised fresh leaves are boiled.
91.	Cyperus rotundus Linn.	Cyperaceae	Roots	Snake bite	Juice taken orally.
			Roots	Asthma	Powder with water taken orally.
92.	Catharanthus roseus (Linn.) G. Don	Apocynaceae	Bark	Excessive or irregular bleeding	Bark paste with cow milk, consumed orally.
93.	Celastrus paniculatus Willd.	Celastraceae	Fruit, seed	Dropsy	Pulp of fruit and seed powder blended with local wine and consumed for seven days.
			Stem bark	Snakebite	Crushed stem bark applied locally.
94.	Ceropegia bulbosa Roxb.	Asclepiadaceae	Tuber	Scorpion sting	Stone dried tuber is rubbed and applied.
95.	Chlorophytum tuberosum Roxb.	Asparagaceae	Root	Sexual debility in man and woman, body heat	Dried fasciculate roots taken orally in form of decoction.
96.	Diospyros melanoxylon Roxb.	Ebenaceae	Stem bark	Malaria	Decoction of stem bark.
97.	Daucus carota Linn.	Apiaceae	Leaves	Migraine	The warmed extract dropped in nostrils and ear three times a day.
			Seeds	Stone in kidney	20 gm seeds of carrot, 20 gm seeds of shalgum, 100 gm radish seed grounded in pastel mortar and given orally with water.
			Seeds	Irregular menstrual cycle	Decoction prepared from carrot, radish and methi

S. No	Name of Plant	Family	Plant Part	Ailment	Mode of administration
					seeds (10 gm each) and Lahori namak boiled in 300 ml water and mixed with black sugar given orally for three to four days.
98.	Dalbergia sissoo	Fabaceae	Leaves	Boils	Paste applied locally.
			Leaves	Gonorrhoea	Decoction mixed with honey taken orally.
			Leaves	Peptic ulcer	Paste taken with sugar.
			Leaves	Dysentery	Paste taken orally.
99.	Datura innoxiaMill .	Solanaceae	Whole plant	Asthma	Smell inhaled through nostrils.
			Roots	Boils	Paste applied locally.
			Roots	Ringworm	Juice applied locally.
			Leaves	Asthma with cough	Powder with goat milk or honey taken orally.
100.	Datura metel Linn.	Solanaceae	Leaves	Asthma	Smoke the dried leaf.
			Leaves	Asthma	Smoke inhaled.
			Roots	Boils	Paste applied locally.
			Seeds	Asthma	Smoked in sulphi.
			Fruits	Malaria	Powder taken orally.
			Dried leaves and seeds	Asthma and Respiratory tract infection.	Two gm powder taken orally with water twice a day for seven days.
101.	Datura stramonium Linn.	Solanaceae	Fruits	Diarrhoea	Fruits given orally for three days.
102.	Desmostachya bipinnata (Linn.) Stapf	Poaceae	Whole plant	U.T.I.	Decoction taken orally.
103.	Desmodium velutinum	Fabaceae	Leaf	Skin diseases, on wound	Crushed leaf paste is applied locally.

S. No	Name of Plant	Family	Plant Part	Ailment	Mode of administration
	(Willd.) DC.				
104.	Dracaena trifasciata	Asparagaceae	Leaf	Cuts, sores, grazes	Leaf sap is applied directly.
105.	Dichrostachys cinerea	Fabaceae	Leaves	Gonorrhea, syphilis	Decoction taken orally.
106.	Enicostemma littorale Acut. Non. Blume	Gentianaceae	Root	Fever	Root is crushed into paste, mixed with curd and taken in fever.
107.	Echinops echinatus Roxb.	Asteraceae	Roots	Protrusion	Decoction orally.
			Roots	Sexual vigor	Boiled with milk and taken orally.
			Roots	Cough and cold	Extract orally.
			Whole plant	Hydrophobia in human beings	Juice of plant mixed with equal quantity of ghee and taken orally.
			Roots	Jaundice	Powder of dried roots mixed with sugar and cardamom taken orally for seven days.
			Roots	Wound healing	Dried powder applied locally.
			Roots	Cough	Powder taken orally twice day.
			Roots	Easy delivery of women	Roots tied up on ponytail of women just before delivery.
108.	Eclipta prostrata Linn.	Asteraceae	Leaves	Uterine bleeding or uterine hemorrhages	Decoction taken orally.
			Whole plant	Burns / cuts	Paste applied locally.
109.	Ephedra foliata	Ephedraceae	Young twigs	Typhoid, Cough and Cold	Decoction taken orally.
			Flower	Protective cure to mosquitoes	Dried powder taken with mattha regularly.

S. No	Name of Plant	Family	Plant Part	Ailment	Mode of administration
110.	Euphorbia hirta Linn.	Euphorbiaceae	Whole plant	Ring worm	Paste applied locally.
			Whole plant	Toothache	Juice used for gargling.
			Whole plant	Leucorrhoea	Decoction orally.
			Leaves	Cough	Decoction given orally with jiggery.
			Stem, leaves andLatex	Piles	Paste of stem and leaves mixed with latex and applied externally on piles for a week.
111.	Euphorbia cordifolia	Euphorbiaceae	Whole plant	Bone fracture setting and rejoining	Mixed with milk and given orally for one month.
112.	Eclipta alba (Linn.) Hassk.	Asteraceae	Root	Liver disorders	Root powder of bhringraj taken orally.
113.	Emblica officinalis Linn.	Phyllanthaceae	Fruit	Anticancer, cold and cough	Fruit consumed directly.
114.	Ficus benghalensis Linn.	Moraceae	Leaves	Diarrhoea	Extract taken orally.
			Leaves	Sprain and blood clotting/thrombosis in muscles	Roasted leaf with honey. Tied on Thrombosis area for two days two or three times at bed time.
			Respiratory root (aerial coir roots)	For birth of male progeny	Five gm coconut mixed with gur (older). Three dose prepared of 3 gm. Prepared tablet is eaten orally at evening from third day of menstrual cycle.
			Latex	Sexual vigor	Milk/latex dropped in Battasa and taken orally daily in morning for fifteen days.

S. No	Name of Plant	Family	Plant Part	Ailment	Mode of administration
			Latex	Piles	Eaten orally to cure piles
			Stem	Spongy gums	Decoction used for gargle
			Latex	Genital disease	Locally applied on organ
			Seeds	Sexual potential	Taken orally with jaggery
			Latex	Diarrhoea and dysentery	Latex poured in batasha and taken orally
			Bark of stem	Leucorrhoea	50 ml decoction mixed with Londhra powder taken orally and vagina washed for seven days
115.	Ficus racemosa Linn.	Moraceae	Stem	Spongy gums	Decoction used for gargle
			Latex	Diarrhoea	Latex taken orally with sugar
			Stem bark	Acne and pimples	Ash based paste used locally with ghee
			Leaves	Vomiting	Given three times orally
			Roots	Stone in urinary bladder (calculus)	Latex taken orally for one month twice a day
			Receptacle	Bleedy vomiting in tuberculosis	Given orally for three days
			Roots	Diabetes	Sap given orally
			Roots	Against habitual abortion	Decoction given orally
			Roots	Measles	20 ml juice of roots. mixed with 5 gm sugar or mishree given orally for three days
116.	Ficus religiosa Linn.	Moraceae	Coir (Stem originated coir)	Male child birth	One gm coir of peepal mixed similar quantity of nag keshar and Shivalingi Seeds taken with cow milk from fourth day to ninth day of menstruation.
			Leaves	Baltod/ abscess	Ghee and sindoor paste applied on leaf and tied

S. No	Name of Plant	Family	Plant Part	Ailment	Mode of administration
					locally
			Root	Asthma upper respiratory infection	Decoction of one kg roots mixed in four liter water and divided in four parts and one fourth liter each of it is taken orally everyday.
			Leaves	Eczema	Fresh leaves chewed for fifteen days.
117.	Grewia abutilifolia Vent. Ex Juss.	Tiliaceae	Leaves	Swollen body part	Warmed leaves are tied on affected part.
118.	Grewia asiatica Acut. non Linn.	Malvaceae	Root bark	Rheumatism	Root bark paste is used by tribals.
119.	Grewia tenax (Forsk.) Fiori	Malvaceae	Plant	Bone fracture and strengthening, tissue healing	Plant preparation taken orally.
120.	Hygrophila spinosa	Acanthaceae	Seeds	Diarrhoea	Seeds are ground to a paste and given with buttermilk.
121.	Holarrhena antidysenterica Linn.	Apocynaceae	Stem bark, stem	Cold, cough, malaria	Extract of bark crushed with Diospyros melanoxylone stem bark extract and taken orally.
			Stem	Heart diseases	Decoction of stem is taken twice a day.
122.	Holoptelea integrifolia (Roxb.) Planch.	Ulmaceae	Leaves	Ringworm	Paste of the fresh leaves, Leaves are crushed with garlic bulblets, mixed with 2-3 drops of lemon juice and applied on the affected area.
123.	Hibiscus rosa-sinensis	Malvaceae	Flowers	Eyesight	Decoction of flowers is used with raw sugar mixed with two gm

S. No	Name of Plant	Family	Plant Part	Ailment	Mode of administration
	Linn.				butterand taken every morning.
124.	Heliotropium indicum Linn.	Boraginaceae	Juice	Wounds, skin ulcers	Extracted juice from the pounded leaves of the plant
125.	Hemidesmus indicus (Linn.) R. Br.	Asclepiadaceae	Roots	Regular tonic	Roots are boiled in water and consumed.
126.	Ichnocarpus frutescens (Linn.) R. Br.	Apocynaceae	Roots	Stone/calculus in kidney and urinary bladder	Powder mixed with water given orally
			Roots	To stimulate appetite and as aphrodisiac	Powder decoction given orally for a week daily in morning
127.	Jatropha gossypifolia Linn.	Euphorbiaceae	Leaves	Eczema and carbuncles (cluster of boils)	Paste applied locally
			Latex	Burns/boils and sores	Applied locally
			Stem	Toothache /weak tooth gums and pyorrhoea	Decoction taken orally or teeth brushed by stem stick
			Roots	Leprosy	Paste locally applied on wounds
			Roots	Inflammation	Paste applied locally
			Seeds	Constipation	Orally as purgative
128.	Jasminum grandiflorum Linn.	Oleaceae	Leaf	Ulcer	Chewed orally
			Leaves	Erectile dysfunction	Fresh leaves and flower's extract used for massage on penis.
			Leaf	Erectile dysfunction	Leaf extract mixed with Brassica juncea seeds and boiled with Sesamum species oil, used externally for massage on sex organs and thigh for erectile dysfunction.

S. No	Name of Plant	Family	Plant Part	Ailment	Mode of administration
			Five organ of the plant	Menstrual disorder	Five organs of chameli (flower, leaf, fruit, bark etc.) given orally.
			Leaf	Crack heel	Extract of leaves applied locally.
			Flower	Eye Cataract	5-6 flowers crushed with mishree and dropped in eye.
129.	Leptadenia pyrotechnica (Forsk.) Decne.	Apocynaceae	Roots and whole plant	Rheumatism	Juice applied locally.
130.	Launaea procumbens (Roxb.)	Asteraceae	Stem	Bleeding	Inner layer of stem is crushed and squeezed and applied.
131.	Lagenaria siceraria (Molina) Standl.	Cucurbitaceae	Fruits	Ear pain	10 gm juice mixed with milk and dropped in ear for three days.
			Fruits	Dysentery and bloody piles	Paste given orally in bed time for three days (each dose of six gm).
			Fruits	Cardiac problem (B.P.) Chest pain	Juice mixed with amla and taken orally.
132.	Lablab-purpureus Linn.	Fabaceae	Seeds	Sciatica Leucorrhoea	Powder mixed with honey given orally.
133.	Lawsonia inermis Linn.	Lythraceae	Leaves	Conjunctivitis	Paste as suppository (inserted).
			Leaves	For curing any pinworm	Paste applied locally.
			Roots	Birth control	Roots soaked in water for twenty four hours and the mediated water is taken orally.
			Seeds	Internal inflammation of body	Paste with cow milk taken orally in the morning.

S. No	Name of Plant	Family	Plant Part	Ailment	Mode of administration
			Bark	Fever and jaundice	Decoction taken orally.
			Seeds	Diarrhoea with blood	Powdered and cold extract mixed with sugar, taken orally.
			Seeds	Infertility (conceiving failure)	100 gm seeds with curd given orally everyday for fifteen days to esure pregnancy.
			Fruits and leaves	As insecticides and body heats (internal)	Paste of fruits and leaves of Lawsonia and Nicotiana tabacum applied locally.
134.	Lantana camara Linn.	Verbenaceae	Whole plant	Bronchitis	Infusion of the whole plant.
			Root	Stomach-ache, vermifuge	Powdered root in milk given to children.
			Whole plant	Leprosy, scabies	Decoctions were applied externally.
135.	Lygodium flexuosum	Lygodiaceae	Rhizomes	Ear-ache	Decoction taken orally.
136.	Leea indica	Vitaceae	Shoot	Sores	Decoction of the shoots taken orally.
137.	Leea macrophylla	Vitaceae	Root, stem	Typhoid	Decoction of stem and root taken orally.
138.	Lactuca serriola	Asteraceae	Plant	Dry cough, rheumatic pain	Plant powder.
139.	Leucaena leucocephala	Fabaceae	Seeds	Stomach, diabetes	Decoction of seeds.
140.	Mangifera indica Linn.	Anacardiaceae	Cotyledons	Diarrhoea	Extract taken orally with processed curd.
			Gums	Scorpion sting	Powder taken orally
			Gums	Chronic dysentery	Gum of this plant mixed with equal amount of its seeds and taken orally in morning.

S. No	Name of Plant	Family	Plant Part	Ailment	Mode of administration
			Leaves	Krack voice	Chewed directly.
			Leaves	Sore throat, Diabetes	Chewed.
			Stem bark	Abortion	Taken orally with bark of Syzigium and Ziziphus.
141.	Momordica dioica Roxb. Ex Willd.	Cucurbitaceae	Leaves	Snake bite	Three gram leaves mixed with leaves of Cissampelos pareira (Kalipath). Decoction given orally 2-3 times a day.
142.	Maytenus emarginata (Willd.)	Celasrtaceae	Leaves	Sore and wound	Paste applied locally.
			Seeds	Allergy, scabies	Oil applied locally.
			Whole plant	Pneumonia/ cold fever	Decoction taken orally.
			Fruits	Scorpion sting	Locally applied.
			Fruits	Blood punities	Taken orally daily in morning.
			Fruits	Constipation indigestion	Used as vegetable or sauce and taken daily with lunch.
143.	Mentha spicata Linn.	Lamiaceae	Whole plant	Scorpion sting	Paste applied topically
			Leaves	Diarrhoea	Paste mixed with curd or lassi or juice and taken orally.
			Whole plant	Cholera	Juice taken orally.
144.	Madhuca indica	Sapotaceae	Seed	Itchy skin	Seed oil is laxative. Oil is applied on skin.
			Stem bark	Diabetes	Decoction of stem bark.
145.	Mimosa pudica	Fabaceae	Leaves and seeds	As a brain tonic	Boiled in buffalo milk and taken orally.
			Roots	Tumor of air in body (Vayugoli)	Roots mixed with Abrus and chitawid roots and cooked in Kadahi, sieved and mix the pahadi long and mustard oil prepare

S. No	Name of Plant	Family	Plant Part	Ailment	Mode of administration
					paste and applied locally for seven days.
146.	Momordica charantia Linn.	Cucurbitaceae	Fruits	Diabetes Type-two	Decoction taken orally or used as vegetable.
			Fruits and leaves	Sexual tonic	Juice boiled and residue made into pills and taken with cow milk.
			Fruits	Jaundice	Crushed and inhaled fresh for three days.
147.	Moringa oleifera Lam.	Moringaceae	Leaves	Dog bite	Paste with lime topically.
			Leaves	Cholera	Extract taken orally.
			Stem bark	Diabetes	Decoction taken orally.
148.	Murraya Koenigii Linn.	Rutaceae	Leaf	Indigestion	For making kari in indigestion.
149.	Musa paradisiaca Linn.	Musaceae	Leaves	Diarrhoea	Fresh leaves are given orally twice a day for three days.
150.	Martynia annua Linn.	Martyniaceae	Leaf	Wounds	Paste.
151.	Nyctanthes arbor-tristis Linn.	Oleaceae	Stem bark	Rheumatic pain	Decoction of stem bark is prepared with the bark of Ricinus communis, taken orally for seven days.
152.	Nerium indicum Mill.	Apocynaceae	Leaves, flower	Abortifacient, Malaria	Leaves and flowers paste with brown sugar and goat milk.
			Whole plant	Scabies, swelling	Decoction is used externally.
153.	Nelumbo nucifera Gaertn.	Nelumbonaceae	Fruits/ buds	Jaundice	Crushed in water and inhaled by nostrils daily for three days.
154.	Nelumbo speciosum	Nelumbonaceae	Pollen	Prevent habitual automatic abortion	Kamal keshar mixed with 10 gm Mullehathi, Decoction given orally with milk daily for nine months.

S. No	Name of Plant	Family	Plant Part	Ailment	Mode of administration
155.	Nerium oleander	Apocynaceae	Seed, leaves	Asthma, heart diseases, cancer	Medicine.
156.	Ocimum basilicum Linn.	Lamiaceae	Leaves	Bee sting	Paste applied locally.
			Leaves	Cold cough and fever	Decoction consumed with honey.
157.	Ocimum sanctum Linn.	Lamiaceae	Inflorescence	Diabetes	Powder taken orally with water.
			Leaves	Cough and cold toothache	Chewed.
			Leaves	Pneumonia	Powder taken orally.
			Leaves	Malaria and intermittent fever	Boiled with tea taken orally.
			Leaves	Ringworm	Extract locally.
			Leaves	Gout/ Gonorrhea	Decoction taken orally
			Seeds	Cancer	White, black, and other types of tulsi mixed and given orally for six months.
			Leaves	Scorpion/ snake and mosquito bite	Juice rubbed locally.
			Seeds	Increasing sperm count	Powder taken orally with cow milk.
			Seeds	Constipation piles	Powder taken orally for one month.
			Seeds	Kidney stone	Powder taken orally.
			Leaves	Heat/sun stroke	Juice of 10 g of leaves mixed with sugar, taken orally thrice a day.
			Roots	Menorrhoea (red discharge)	Powder or roots put on the piper betal leaf and chewed orally.
			Leaves	U.T.I. (inflammation in genital organ)	Chewed orally.
			Roots	Restarting menstrual cycle which suddenly	Fifty gm roots of tulsi taken and prepare decoction, one dose given

S. No	Name of Plant	Family	Plant Part	Ailment	Mode of administration
				stopped	orally before lunch for seven days.
158.	Oxalis corniculata Linn.	Oxalidaceae	Leaves	Mouth fresh	Leaves chewed.
			Leaves	Diarrhoea	Leaf extract taken orally.
			Leaves	Prolapse of Anus	Extract mixed in ghee and boiled, taken orally.
			Bark of stem	Red/white discharge	Powder of bark mixed with sugar (mishree) and taken orally with processed curd (mattha).
			Stem/twigs	Crack heel	Maida lakdi and Lokwan paste applied locally.
159.	Papaver somniferum Linn.	Papaveraceae	Latex	Sujan (Arthritis with swelling) in knee and ankles	10 gm opium latex, 10 gm Desi long, 10 gm Pahadi long, 200g, sesame oil, 200 gm cuscuta, mixed and roasted in oil, prepare ointment for massage daily.
			Fruits	Vomiting	Roasted and taken on wet cloth and prepare paste/powder, taken orally mixed with honey.
			Latex	Cooling purpose	Latex half gm and two pieces of draksha, insert inside the bulb or onion and roasted half and eaten orally.
160.	Pedalium murex Linn.	Pedaliaceae	Seeds	Sexual weakness impotency	One gm each of Gokharu, gourakh mundi, Bophali, Kharenti seeds, safed musli salim panja, Asparagus taken and given orally with butter and mishree or milk in morning daily.
			Whole	Sexual vigor	Juice taken orally.

S. No	Name of Plant	Family	Plant Part	Ailment	Mode of administration
			plant		
			Fruits	Weakness	Mixed with wheat flour sugar and ghee and taken orally.
			Plant	Loss of appetite	Extract taken orally.
			Fruits	Syphilis	Fruits of this plant mixed with Abrus (Ratti) and decoction prepared and taken orally.
			Leaves	Gonorrhea	Sap taken orally.
			Seeds	Dysentery	Dried or fresh seed grounded with water and mixed with chhach taken orally.
			Seeds	Infertility and sexual incompatibility	50 gm seeds for each dose taken and mixed with almond, grapes and sugar taken orally in morning with cow milk.
			Seeds or Whole plant	UTI, diabetes mellitus, Prostration inflammation erectile dysfunction	Seeds mixed with sugar (mishri) and taken orally in morning or mixed with flower and make laddu that are eaten orally in morning.
161.	Phoenix sylvestris (Linn.) Roxb.	Arecaceae	Seeds	Pain and swellings	Powder and paste applied locally
			Fruits	Arthritis	Given orally regularly in winter season.
			Leaves	Swelling of testis	Paste applied locally.
			Fruits	Weakness	Boiled in milk and taken orally.
162.	Pithecellobium dulce (Roxb.) Benth.	Fabaceae	Bark, twig	Tooth medicine	Extract of bark and twig.

S. No	Name of Plant	Family	Plant Part	Ailment	Mode of administration
163.	Phyllanthus fraternus var niruriAcute. Pl. non Linn.	Euphorbiaceae	Leaves	Skin diseases wound and fractures	Paste applied locally.
			Leaves	Intermittent fever	Decoction taken orally.
			Roots	Jaundice	Decoction taken orally.
			Whole plant	Constipation	Crushed the whole plant and filtered juice taken orally for fifteen days.
			Whole plant	Diabetes	Extract given orally for relaxation.
			Whole plant	Hepatitis	Extract given orally for fifteen days.
164.	Phyllanthus emblica Linn.	Euphorbiaceae	Fruits	Leucorrhoea	Soaked in honey and taken orally.
			Fruits	Diabetes	Juice mixed with juice of Momordica charantia and taken orally.
			Fruits	Rheumatism	Powder mixed with jaggery and taken orally.
			Fruits	Diarrhoea	Powder taken orally.
			Seeds	Syphilis	Orally with honey.
			Seeds	Leucorrhoea	Grounded with water and sieved, honey and sugar mixed in decoction and taken orally.
165.	Piper longum Linn.	Piperaceae	Roots	Loss of sleeping (Insomnia)	70 gm powder of roots mixed with black sugar and taken orally in morning for seven days (10-10 gm daily).
			Fruits	Moderate fever / Hadjuri	Fruits socked in mattha over night and then dried. Powder mixed with gur and given orally for three days.
			Fruits	Galactagogue	Fruit powder mixed with

S. No	Name of Plant	Family	Plant Part	Ailment	Mode of administration
					satawar powder and given orally.
			Roots	Heart disease	Root powder and choti ilaichi taken in equal quantity and grounded and mixed with cow ghee given orally.
			Fruits	Sciatica	Fruits powder mixed with ginger powder and decoction prepared in sesamum oil and is used for massage.
166.	Piper nigrum Linn.	Piperaceae	Seeds	Common cold and cough	10 gm pepper roasted and powdered and given orally with hot milk during bed time for two days.
			Seeds	Dry cough	15 gm piper mixed with desi ghee and mishree and taken orally for three days.
167.	Prosopis cineraria (Linn.) Druce	Fabaceae	Bark	Leprosy, dysentery, asthma, leucoderma	Making medicines.
168.	Plumbago zeylanica Linn.	Plumbaginaceae	Leaves	Antidote for snake bite	Juice of 5-10 leaves is taken orally.
			Root	Ringworms	Paste of fresh root is applied locally once a day for two days.
			Root bark	Rashes, hemorrhage	Sap of root barkapplied locally.
169.	Physalis angulata	Solanaceae	Leaves	Lithiasis (stony concretions), stomachache, colic	Leaves applied.

S. No	Name of Plant	Family	Plant Part	Ailment	Mode of administration
170.	Pongamia pinnata Linn.	Fabaceae	Root, stem bark	Boils, rheumatic pain	Crushed root and stem bark is boiled in mustard oil and this medicated oil is used to massage the body.
171.	Rauvolfia serpentina	Apocynaceae	Root	Rheumatism	Decoction.
172.	Ricinus communis Linn.	Euphorbiaceae	Leaves	Reduce menses pain (painful mens)	Juice with mustard oil applied locally.
			Leaves	Guinea worm	Roasted half leaves tied locally.
			Leaves	Sprain and fractures	Roasted half leaves applied hot locally.
			Leaves	Abdominal pain/gastric disturbance	Tied hot locally on abdomen.
			Leaves	Jaundice	Given orally for three days.
			Seeds	Birth control	Orally with water during menstruation.
			Seeds	Birth control	Oil locally applied on sex organ at the time of coitus.
			Seeds	Dyspepsia	Pulp taken orally with milk
			Oil	Gastritis	One spoon castor oil mixed with fresh curd and given orally 2-2 spoons at interval of two hours for fast relief.
173.	Saccharum bengalense var munjaRetz.	Poaceae	Roots	As aphrodisiac and general tonic	Decoction given orally regularly for fifteen days.
174.	Syzygium heyneanum (Duthie)	Myrtaceae	Bark	Wound	Paste of bark is applied locally.

S. No	Name of Plant	Family	Plant Part	Ailment	Mode of administration
	Wall. Ex Gamble				
175.	Saccharum officinarum Linn.	Poaceae	Juice	Headache	Make the desi gur by process of juice and consumed, four piece of gur thrown in four different directions.
			Roots	Birth control	Root powder taken orally with cow milk thrice for three days.
176.	Salvadora oleoides	Salvadoraceae	Roots	Dental ailments	Chewed.
177.	Salvadora persica	Salvadoraceae	Twigs	Pyorrhoea	Used as brush.
			Roots	Tenderness	Paste mixed with processed curd and applied locally.
			Roots	Abortifacient	Roots powder mixed with water and given 2-3 times orally.
			Roots	Tenderness	Paste of roots mixed with curd and applied locally.
			Roots	Abortion stimulating	Root powder given orally with water, 2-3 times in a day.
			Leaves	Appetite loss	Paste taken orally with honey.
			Branches	Pyorrhoea	Used for dental brush.
178.	Santalum alba	Santalaceae	Heartwood	Acne/pimple	Two spoon each of neem juice, gulab jal, multani mitti, chandan and two drops piperment oil. Mixed and applied as face pack.
179.	Sesbania sesban (Linn.)Merr.	Fabaceae	Leaves	Diabetes	Paste taken orally.
18	Solanum	Solanaceae	Leaves	Whooping cough	Extract taken orally twice

S. No	Name of Plant	Family	Plant Part	Ailment	Mode of administration
0.	nigrum Linn.				a day.
			Leaves	Dog bite	Paste applied locally for three days.
			Roots	Fever	Decoction taken orally.
			Roots	Cold, cough and asthma	Decoction taken orally regularly for fifteen days.
			Fruits	Diarrhoea, Fever, Hydrophobia	Taken orally or chewed.
			Seeds	Aphrodisiac use	Taken orally regularly for one month.
			Whole plant	Liver tonic	Decoction orally.
181.	Solanum surattense	Solanaceae	Fruits	Cough	Chewed.
			Fruits	Toothache and jaw ache	Chewed directly.
			Flowers	Diarrhoea	Given orally directly.
			Leaves	Rheumatism	Given orally to patient.
			Roots	Cough asthma and fever	Decoction given orally for five days.
			Seeds	Irregular menstrual cycle	Eaten orally with water.
			Whole plant	Syphilis	Paste applied locally.
			Whole plant	Piles	Locally applied.
			Seeds	Cough, URI	Ten gm seeds half roasted on tawa and mixed ten gm jagerry.Ten tablets of 2-2 gm made and taken orally twice a day for three days.
			Seeds	Germ in tooth and toothache	Vapour used to kill the germ mustard oil and seeds of Solanum roasted and dropped in ear.
182.	Solanum xanthocarp	Solanaceae	Fruits and seeds	Skin infection	Make halwa of flour and mixed fruits seeds which

S. No	Name of Plant	Family	Plant Part	Ailment	Mode of administration
	um Linn.				are roasted in ghee and given orally.
			Roots	Thorn inside body	Roots and gur mixed up to make ten tablets of 2-2 gm that are taken orally for three days.
			Fruits	Ring worm	Fruits cuts in two pieces and rubbed locally.
183.	Sonchus oleraceus Linn.	Asteraceae	Whole plant	Piles	Paste applied topically for a week.
184.	Syzygium cumini Linn.	Myrtaceae	Stem bark	Dysentery	Crushed, given orally thrice a day.
			Leaves	Diabetes	Leaves chewed daily in morning and evening.
			Seeds	Diabetes	Powdered seeds mixed with triphala and leaves of Gymnema sylvestre and Mangifera indica taken orally twice a day for relief.
			Leaves	Diarrhoea/ Loose motion	Paste mixed with fresh curd and given orally.
			Bark	Loose motion and dysentery	Bark mixed with curd and given orally.
185.	Sida glutinosa	Malvaceae	Leaves	Inflammation	Paste.
186.	Tamarindus indica Linn.	Fabaceae	Seeds	Scorpion stings	Rubbed on stone with water and sap used locally.
			Stem bark	Diarrhoea	Stem bark decoction is given twice a day.
187.	Tephrosia purpurea	Fabaceae	Root	Toothache	Root is chewed slowly twice a day.
			Leaves	Pile bleeding	Paste of leaves is made with leaves of Cannabis sativa and applied locally.

S. No	Name of Plant	Family	Plant Part	Ailment	Mode of administration
188.	Trapa natans Linn.	Lythraceae	Seed	Leucorrhoea	Seed powder is used to prepare laddoo that are eaten by women.
189.	Tamarix aphylla	Tamaricaceae	Roots and stem bark	Burns, ulcers and wound	Paste applied locally.
190.	Thevetia peruviana (Pers.) Merr.	Apocynaceae	Roots	Eczema and other skin disease	Roots roasted with oil of sesamum to make decoction, and paste applied locally with besan.
191.	Tecomella undulata (Sm.) Seem	Bignoniaceae	Bark	Syphilis	Chewed directly for seven days.
			Bark	Birth control	Chewed directly for three days.
			Bark	Leucorrhoea	Powder mixed with honey and sugar and taken orally with milk.
192.	Terminalia arjuna (Roxb. Ex DC.)	Combretaceae	Stem bark	Heart problems obesity, night fall, loss of appetite, moderate fever, dysentery, wound healing etc.	Make (Arjun ksheera) and taken orally for three months regularly for relaxation from mentioned disease. Method to prepare the ksheera, one cup milk of cow, one cup of water, ten gm Arjun powder, decoction preparedmixed with sugarand taken orally.
			Stem bark	Chronic acute wound healing	Decoction used for washing the wound daily for fifteen days.
			Stem bark	Moderate fever	Arjuna ksheera Decoction taken orally in morning for fifteen days.
			Stem bark	Dysentery	Used the goat milk to prepare Arjuna ksheera and taken orally twice a day, powder also taken

S. No	Name of Plant	Family	Plant Part	Ailment	Mode of administration
					twice a day for fast relief.
193.	Tinospora cordifolia (Willd.) Miers	Menispermaceae	Leaves	Moderate fever (Had juri)	Five leaves taken and mixed with two gm Piper longum and seeds of black pepper mixed and taken two to three times orally.
			Stem	Piles	Given orally.
			Whole plant	Galactagogue	Crushed whole plant given orally for seven days.
			Seeds	Protrusion of uterus	Half kg seeds half kg pulse of gram soaked in water overnight and mixed with misri and given orally
			Leaves	Leucorrhoea	Decoction taken orally.
			Stem	Piles	Extract mixed with Triphala, turmeric and chirayata powder and given orally.
			Stem	Fever, Asthma, Diabetes	Decoction taken orally.
			Stem	Painful lactation	Extract or paste rubbed locally on mother's breast two to three times for three days.
194.	Trachyspermum ammi Linn.	Apiaceae	Seeds	Scanty menstrual cycle	Powder orally taken two to three times a day.
			Seeds	Gastrointestinal Disorders	Given orally in morning before breakfast for three days.
			Seeds	Common cold, influenza	20 gm seeds roasted on Tawa and taken on wet cloths, then roasted seeds are inhaled.
195.	Tribulus terrestris	Zygophyllaceae	Fruits	Impotency	Powder mixed with seeds of Sesamum indicum taken

S. No	Name of Plant	Family	Plant Part	Ailment	Mode of administration
	Linn.				orally with milk.
			Fruits	Whooping cough	Powder taken orally.
			Plant	Sexual debility	Decoction mixed with almond and honey taken orally with cow milk.
196.	Tridax procumbens Linn.	Asteraceae	Leaves	Bleeding from wounds and cuts	Juice applied locally.
			Leaves	Menorrhoea	Extract with mattha given orally for three days.
			Leaves	Pus in ear	Extract dropped in ear for three days.
			Stem	Tuberculosis	Extract mixed with juice of Ocimum sanctum given orally for three months.
			Stem	Gonorrhea	Extract mixed with honey and sugar and given orally for five days.
			Stem	Cancer	Extract mixed with 200 ml wheat grass extract, Ocimum sanctum extract and extract of Azadirachta indica given orally for three months.
			Leaf and Whole plant	Cuts and bleeding / ear pus, burns and wounds	Juice dropped locally.
197.	Trigonella foenum-graecum Linn.	Fabaceae	Leaves	Leucoderma	Locally applied.
			Leaves	Swelling	Paste applied locally.
			Seeds	Diabetes	Paste taken orally.
			Seeds	Stomachache	Taken orally with water.
			Seeds	Dysentery, diarrhoea dyspepsia	Given orally with water or mattha in morning.
			Seeds	Hyper and hypo tension (B.P.)	50 gm seed soaked in water over night and eaten daily for fifteen

S. No	Name of Plant	Family	Plant Part	Ailment	Mode of administration
					days.
			Leaves	Irregular menstrual cycle	Dried and pounded leaves are applied locally.
			Roots	Asthma and fever	Decoction taken orally.
			Roots	Liver function	Powder taken orally.
			Roots	Tuberculosis	Decoction with black pepper and honey taken orally.
			Roots	Cold and cough	Powder or decoction taken orally.
			Roots	Diabetes	Powder of nagori ashwagandha one spoon mixed with chardraprabha vati tablespoon of Ayurvedic pharmacy one-one tablespoon taken after meal, silajeet ras prapati taken with honey and cow milk daily.
			Roots	Rheumatism, knee pain, jaundice, U.T.I	Powder of roots mixed with honey and taken orally.
			Fruits	Inflammation	Two gm fruit mixed with munakka and taken orally.
			Fruits	Tenderness	Used as vegetable and roasted half leaves are tied on affected part.
			Leaves	Armpit abscess/ boils	Roasted leaves tied on arm pit and boils locally.
			Leaves	Abscess boil	Paste of leaves mixed with cow urine and used locally on burnt area.
			Roots/leaves	Menorrhoea	Roots powder given orally with rice Dhowan (Rice water).

S. No	Name of Plant	Family	Plant Part	Ailment	Mode of administration
			Leaves or latex	Boils/ abscess	Latex applied locally and leaves tied on boil.
			Whole plant	Leucorrhoea	Paste mixed with milk and mishree given orally empty stomach for three days.
			Whole plant	Tenderness	Mixed with water and given orally.
			Stem and leaves	Toothache	Extract dropped in ear.
			Whole plant	Ricketts (dry diseases of child)	Decoction used for bath.
			Leaves	Antifungal, Ring worm	Rubbed on affected part.
			Latex	Wounds, Abscess, burns etc.	Locally and new born leaves tied with warmghee.
			Seeds	Restarting the menstrual cycle in case of complications.	Ten gm til mixed with equal amount of roots of piper layer and ghee to make a decoction which is consumed orally.
			Leaves	Mumps	Roasted half leaves tied on cheek.
198.	Tabernaemontana undulata	Apocynaceae	Bark, branch	Syphilis	Chewed.
199.	Thespesia populnea	Malvaceae	Leaf	Skin disease	Leaf sap used externally.
200.	Wrightia tinctoria (Roxb.) R.Br.	Apocynaceae	Leaves	Scorpion bite	Chewed orally.
			Leaves	Toothache	Paste as tooth powder.
			Roots and stem	Abdominalpain	Extract taken orally
201.	Withania somnifera Linn.	Solanaceae	Root	Rheumatism	One tea spoon full of root powder is given with 250 ml milk twice a day.
			Leaves	Tonsils, pimples	Mature leaves are

S. No	Name of Plant	Family	Plant Part	Ailment	Mode of administration
					smeared with ghee or oil warmed and tide locally on boils.
			Leaf	Skin diseases	Leaf juice is applied locally.
			Root	Rheumatic pain	Root powder is used to make ladoo and given orally.
202.	Vitex peduncularis	Lamiaceae	Root	Jaundice	Powder.
203.	Vitex negundo Linn.	Lamiaceae	Leaf	Conjunctivitis	Leaf extract is poured in the eyes.
			Leaves	Rheumatism	Leaves are boiled in earthen pot and used orally twice a day.
204.	Xanthium strumarium Linn.	Asteraceae	Leaves	Skin disease/ Excema	Powder mixed with mustard oil and applied topically.
			Fruits	Cough and cold	Leaf of Psidium guajava mixed with fruits of Xanthium to prepare decoction that is taken orally.
205.	Zingiber officinale Roscoe	Zingiberaceae	Rhizome	Wet cough	Ten gm of ginger mixed with ten gm black sugar, sendha namak to make decoction that is given orally twice a day (one spoon).
			Rhizome	Reduce labor pain	Powder taken orally with warm milk.
206.	Ziziphus mauritiana Lam.	Rhamnaceae	Leaves	Diabetes	Chewed or powder taken orally with water.
			Roots	Fever	Decoction given orally.
			Root bark	Abortion	Decoction given orally.
			Leaves	Skin diseases boils and scabies	Paste applied locally on affected parts of body.

S. No	Name of Plant	Family	Plant Part	Ailment	Mode of administration
207.	Ziziphus nummularia (Brum.f.) Wight & Arn.	Rhamnaceae	Roots	Expulsion of placenta	Decoction of root given orally.

Summary

Forests have been the primary source for collecting materials for the practice of Ethnomedicine. The evidence of the use of plants for medicine is seen in the fossil record of the Middle Paleolithic period around 60,000 years ago. Tribes in China, India, Romania, and Africa have practiced Ethnomedicine for ages. Ethnomedicine refers to natural practices of healing and treating ailments and diseases using various local practices made of wild plant and animal products. According to some studies, about 75-90% of rural populations rely on traditional medicines for their health care system.

Glossary

1. Ethnomedicinal- Plants used by tribal people in curing health ailments.

2. Ethnoveterinary- Plants used by tribal people in curing health ailments of cattle.

3. Poultices- A soft substance that you spread on a cloth and put on the skin to reduce pain or swelling

4. Decoction- A concentrated liquor resulting from heating or boiling a substance, especially a medicinal preparation made from a plant.

5. Orally- By or done to the mouth.

6. Locally- In a particular place or area

Exercise: Answer the following questions.

1. What is Ethnobotany?
2. Which plant is commonly used in traditional medicine in Rajasthan?
3. What is the significance of Ethnomedicinal plants?
4. Which of the plants is used to treat diabetes in traditional medicine?
5. What is the importance of documenting Ethnomedicinal knowledge?
6. Which plant is used to treat skin conditions in traditional medicine?
7. What is the role of local communities in conserving Ethnomedicinal plants?
8. Which of the following plants is used to treat respiratory issues in traditional medicine?
9. What is the significance of Ethnomedicinal plants in primary healthcare?
10. Which plant is used to treat digestive issues in traditional medicine?

Test Paper

Chapter-5
Medicinal Plants of Rajasthan

Time Duration: 30 minutes **Maximum Marks: 25**

General Instructions:

1. The question paper consists of 12 questions and is divided into four sections, A, B, C and D.
2. All questions are compulsory.
3. Section A comprises MCQS, carrying one marks each.
4. Section B comprises VSAQS (very short answers questions) carrying three marks each.
5. Section C comprises SAQS (short questions) carrying three marks each.
6. Section D comprises long questions carrying six marks each.

SECTION-A

Exercise 1. Multiple choice questions: **(1*4=4)**

1. Which plant is used for curing heart disease?
(a) Arjuna
(b) Cinchona
(c) Tulsi
(d) Neem
2. Ethnobotany refers to relation between
(a) People and plants
(b) People and Water
(c) People and air
(d) None of the above
3. Sacred groves consist of
(a) Sacred plants
(b) Water
(c) Land
(d) All the above
4. Veterinary word is used for
(a) Plants
(b) Cattle
(c) Land
(d) All the above

Answer: 1. (a) 2. (a) 3. (a) 4. (b)

SECTION-B

Exercise 2: Very Short answer type questions **(3*3=9)**

1. What is meant by Ethnomedicine?
2. What are Ethnoveterinary plants?
3. Give examples of some Ethnomedicinal plants of Rajasthan.

SECTION-C

Exercise 3: Short answers type questions **(3*2=6)**

1. What is the most common mode of administration of Ethnomedicinal plants?
2. What parts of the plants are responsible for the medicinal potential of plants?

SECTION-D

Exercise 4: Long answers type questions **(6*1=6)**

1. Give a comprehensive account of the Ethnomedicinal and Ethnoveterinary plants of Rajasthan used by the tribal communities.

Online Resources

	Online Practice Paper	Online Video Link
Link	https://forms.gle/UfsCWa57xCTs1zEQ6	https://www.youtube.com/playlist?list=PLzlh_31VWEFoQwlIptXtKeCjb3RXwcUxn
QR Code		

Chapter 6: Ethnobotanical uses of Plants of Rajasthan

In this chapter, you will learn about:

Ethnobotanical uses of Plants of Rajasthan: Their Description as food sources, including wild edibles and cultivated crops, as timber, agricultural implements, etc., used by the tribal communities of Rajasthan.

Plants as Food and in Agriculture: importance

The plant species grown as crops in the district by tribals of the district are – Brassica campestris, Capsicum annuum, Cicer arietenum, Cucurbita maxima, Cyamopsis tetragonoloba, Eruca sativa, Hordeum vulgare, Luffa acutangula, Pennisetum americanum, Sesamum indicum, Trigonella foenum-graecum, Triticum aestivum, Vigna aconitifolia and Zea mays. Some wild plants used as a sourcc of food by tribals are Acacia nilotica subsp. indica, Aegle marmelos, Albizia lebbeck, Aloe barbadensis, Capparis decidua, Prosopis cineraria and Zizyphus spp.

Famine Food

Plant species utilised as famine food in the area are Acacia nilotica, Achyranthes aspera, Capparis decidua, Cenchrus biflorus, Citrullus colocynthis, Cyperus rotundus, Ficus spp., Pedalium murex, Prosopis cineraria, Salvadora persica, Tribulus terrestris, etc.

Fodder

Many plant species are known to be utilised as sources of fodder for cattle, including cows, horses, buffalo, camels, goats, sheep, and donkeys, which are among the common animals kept by both tribal and non-tribal people in the area. The primary fodder sources include annual and perennial grasses, sedges, herbs, shrubs, and trees.

Plants used in field fencing

Although some plants and trees are not primarily used for field fences, certain trees and shrubs are allowed to grow within or in the vicinity of the fences. Their branches are used in forming dense dead barricades in the form of fencing. Some of their properties, such as their spiny or bushy nature and sand-binding ability, are similar to those of plants grown primarily for fencing. Common plant species used for field fencing by the tribals of the area are Ailanthus excelsa, Leptadenia pyrotechnica, and Acacia nilotica.

Plants providing edible oil, nonedible oil, gums & resins, tannin and detergents in the district

Plant species used as source of edible oil are Brassica campestris, Eruca sativa, Sesamum indicum, Brassicu juncea, Brassica nigra. Plant species used as a source of nonedible oil by the tribals are Azadirachta indica, Brassica juncea, Chenopodium alba, and Musa paradisiaca. Acacia nilotica, Acacia Senegal, Acacia catechu, Azadirachta indica, Ficus benghalensis, Ficus racemosa, and Prosopis cineraria are the primary plant species used by the tribals of the area as sources of gums and resins. Acacia catechu, Anogeissus pendula, and Acacia nilotica are the main plant species utilised for tannins. Plant species are utilised as a source of detergents by the tribal women in the area, Acacia nilotica, Acacia Senegal and Azadirachta indica

Fibre plants

The plant fibre plays an important role in fulfilling various human needs and is a part of their basic requirements. The tribals mainly use fibres, viz. Saccharum bengalense and Crotolaria juncea.

Timber wood

Acacia nilotica subsp. Indica, Anogeissus pendula, Capparis decidua, Dalbergia sissoo, Dendrocalamus strictus, Holoptelea integrifolia, Ficus spp., Prosopis cineraria, and Leptadenia pyrotechnica are common plant species used as timber.

Fuel wood

Capparis decidua, Cajanus Cajan, Azadirachta indica, Anogeissus pendula, Acacia senegal, Ipomoea fistulosa, Pennisetum typhoedium, Ailanthus excelsa, Acacia nilotica, Prosopis juliflora, Prosopis cineraria, Salvadora oleoides, and Zizyphus mauritiana are used for fuel.

Plants used for shelter and agricultural equipment

Shelter

Plants used in making huts and shelters are – Acacia nilotica, Anogeissus pendula, Calotropis procera, Pennisetum americanum, Prosopis cineraria, Prosopis juliflora, Saccharum bengalense, Cajanus cajan, Saccharum spontaneum, and Zizyphus nummularia.

Agricultural equipment

Acacia nilotica, Anogeissus pendula, and Prosopis cineraria are utilised for making agricultural equipment.

Ethnomusicology

Musical instruments like Bansuri, Chang, Dholak, Dhol, Nagada, Poongi, Ravanhattha, Ghera, Duff, Harmonium, Khartal, Sarangi are made from plant sps. Such as – Acacia nilotica subsp. indica, Ailanthus excelsa, Argemone maxicana, Azadirachta indica, Dalbergia sissoo, Dedrocalamus strictus, Mangifera indica.

Plants used as Intoxificants and Masticatories

Many species of Angiospermic plants are used as intoxicants and masticatories by the indigenous people of the district. Plants used as intoxicants and masticatories are Acacia nilotica, Balanites aegyptiaca, Cordia gharaf, Datura innoxia, Datura metel, Calotropis procera, Nicotiana tabacum, Tamarindus indica and Zizyphus nummularia.

Plants used in Ceremonies

The various plants found to be significant in ceremonies in tribal lives are Acacia nilotica, Hordeum vulgare, Lawsonia inermis, Oryza sativa, Pennisetum americanum, Sesamum indicum, Cicer aeritinum, Brassica juncea, Vigna mungo, Vigna radiata, Mangifera indica, Ficus religiosa, Cyanodon dactylon, Saccharum spontaneum, Acacia catechu, Piper betel, Musa paradisiaca, Saccharum officinarum and Ziziphus nummularia.

Summary

In recent years, food security has been a pressing issue in our country. Recent studies have found that wild edible food plants can play a crucial role in the nutrition and income generation of tribal communities. The consumption of wild food plants has been. It continues to be underestimated, and research, particularly concerning the socio–economic, cultural, traditional, and nutritional aspects of wild food plants, needs adequate attention. These wild resources are closely tied to the socio-economic development of rural and tribal peoples, which helps meet their day-to-day needs. Furthermore, there is limited scientific information on the distribution and consumption patterns of wild foods among different communities across landscapes. The older people of tribal communities possess adequate knowledge about wild food plants, whereas the younger generation shows little to no interest in learning about them. There is always a fear of losing invaluable traditional knowledge if it fails to pass from one generation to the next. There is now a growing awareness of the value of such traditional knowledge and a recognition of the urgent need to document it, concomitant with efforts to preserve the natural forest resources that are home to rich floral and faunal diversity. Steps need to be taken for the cultivation and utilisation of important wild edibles in agroforestry systems, which will be a significant factor in the restoration of tradition.

Glossary

1. Agriculture produces commodities which maintain life, including food, fibre, forest products, horticultural crops, and their related services

2. Fodder: something fed to domestic animals; especially: coarse food for cattle, horses, or sheep

3. Famine is a lack of food over a long period in a large area that can cause the death of many people

4. Ethnobotany- Ethnobotany is the study of how people of a particular culture and region make use of indigenous (native) plants

Exercise: Answer the following questions

1. What is the Ethnobotanical use of the plant Aloe vera?
2. Which plant is used by indigenous communities as famine food?
3. What is the traditional use of Turmeric (Curcuma longa)?
4. Which plant is used by local communities to make furniture?
5. What is the Ethnobotanical use of Neem (Azadirachta indica)?
6. Which plant is used by indigenous communities to make agricultural implements?
7. What is the traditional use of the plant Ginger (Zingiber officinale)?
8. Which plant is used by local communities to treat wounds?
9. What is the Ethnobotanical use of the plant Ashwagandha (Withania somnifera)?
10. Which plant is used by indigenous communities to treat fever?

Test Paper

Chapter 6

Ethnobotanical Uses of Plants of Rajasthan

Time Duration: 30 Minutes **Maximum Marks: 25**

General Instructions:

1. The question paper consists of 12 questions and is divided into four sections, A, B, C and D.
2. All questions are compulsory.
3. Section A comprises MCQS, carrying one mark each.
4. Section B comprises VSAQS (very short answer questions) carrying three marks each.
5. Section C comprises SAQS (short questions) carrying three marks each.
6. Section D comprises long questions carrying six marks each.

SECTION-A

Exercise 1. Multiple choice questions: **(1*4=4)**

1. The word herb is derived from-
(a) Herbarium
(b) Herba
(c) Herbeum
(d) Harba
2. Identification test of herbs includes.
(a) Macroscopic character
(b) Chemical reaction
(c) Microscopic character
(d) All the above
3. Herbal drug preparations are prepared by .....
(a) Decoction
(b) Infusion
(c) Maceration
(d) All the above
4. The study of the use of medicinal plants is known as.....
(a) Herbarium
(b) Pharmacognosy
(c) Herbalism
(d) None of the above

Answer: 1. (b), 2. (d), 3. (d), 4. (d)

SECTION-B

Exercise 2. Very Short answer type questions **(3*3=9)**

1. Give examples of plans used in Ceremonies.

2. Give examples of plants used as famine food.
3. Give examples of plants used as fodder.

SECTION-C

Exercise 3. Short answer type questions- (3*2=6)

1. Discuss briefly the importance of plants used in festivals and ceremonies by the tribals.
2. Explain the importance of plants used during famine times.

SECTION-D

Exercise 4. Long answer type questions (6*1=6)

1. What is the role of plants in the life of tribal people of Rajasthan?

Online Resources

	Online Practice Paper	Online Video Link
Link	https://forms.gle/f7CFyU9FAdNbtPhv8	https://www.youtube.com/playlist?list=PLzlh_31VWEFqfdQoMSOLS4z4Uit1DFlp0
QR Code		

Chapter 7: Ethnoveterinary Practices of Rajasthan

In this chapter, you will learn about:

1. **Ethnoveterinary Practices in Rajasthan by the Tribal People.**
2. **Documentation of traditional practices used for animal healthcare**

Plants have long been essential to human survival. Ethnobotany is the branch that illustrates the direct interaction between man and the plants. Ethnomedicinal plants offer a wealth of valuable organic and inorganic compounds that can be utilised to treat a range of diseases. India is a densely populated developing country that is one of the world's most significant biodiversity hotspots, with a diverse plant population. Africa has the largest tribal population in the world, followed by India. India is one of the world's 12 megadiverse countries, and its diverse flora and fauna can be attributed to its various climate zones. India is home to approximately 45,000 plant and 81,000 animal species, accounting for 8% of the world's biodiversity. It is a place with a distinct system of ancient medicinal practices, including Siddha, Ayurveda, and Unani. Surprisingly, the rural sector contributes more to biodiversity than urban areas in the country, where modern health facilities are often insufficient, forcing people to rely on traditional medical techniques and indigenous plants as herbal treatments. Traditional knowledge refers to the knowledge gained by indigenous people through experience and adaptation to their local cultures and environments. In 1895, Hershberger referred to this traditional knowledge as Ethnobotany. The study of the interaction between humans and plants is known as Ethnobotany, which derives from the words "Ethno" (meaning "people") and "botany" (meaning "plant study").

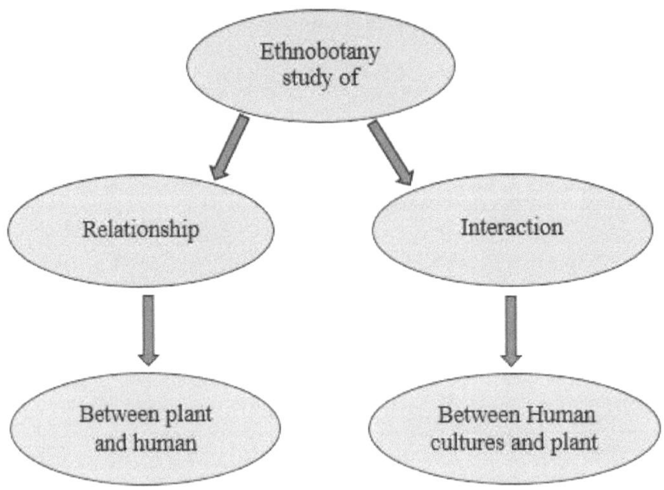

Figure: 1

One subfield of Ethnobiology is regarded as Ethnobotany. Ethnobotany investigates the intricate connections (and applications) between plants and societies. Ethnobotany studies how plants have been used, managed, and perceived in human civilisations. This covers how plants are utilised for building materials, tools, currency, clothing, rituals, social life, and music, as well as for food, medicine, cosmetics, dyes, and textiles. Ethnobotany, a multidisciplinary study, examines the relationship between humans and plants. In addition to being used for food, clothing, and shelter, plants are also utilised in religious ceremonies, as decorative elements, and for medicinal purposes in human societies.

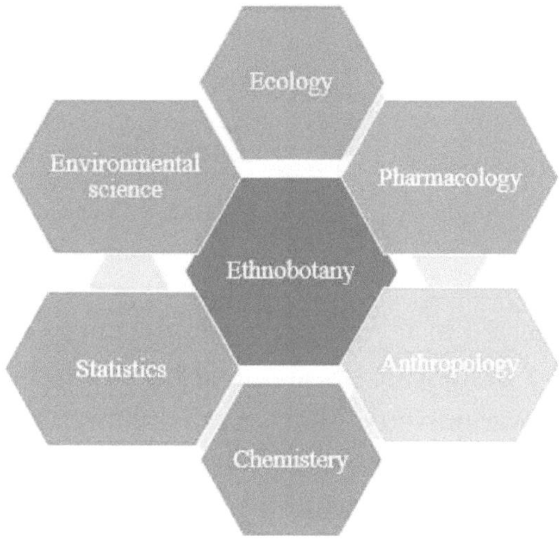

Figure 2: The discipline of Ethnobotany and a few of its related fields

The tribal people were interviewed with the help of a structured questionnaire about the use of plans in curing the health ailments of the cattle. The method adopted for identification is illustrated in the graphic representation.

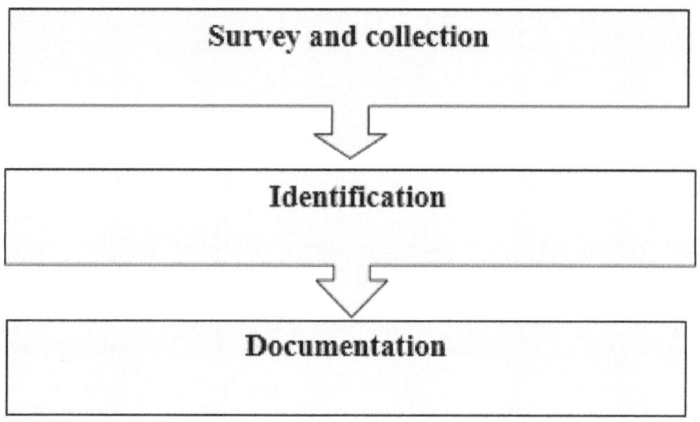

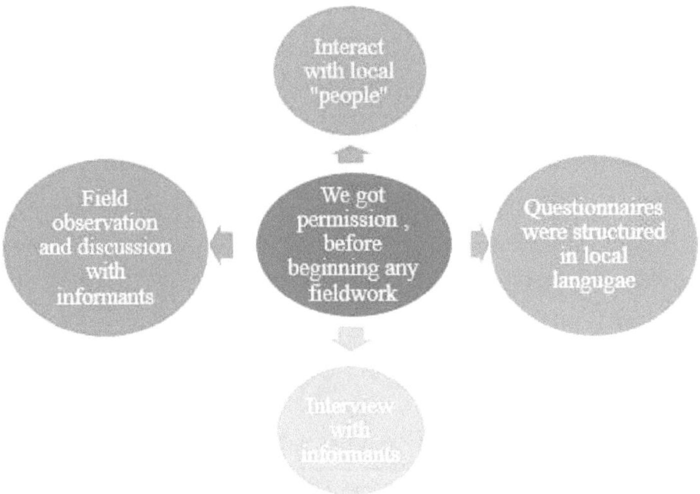

Figure: 3

Field expeditions were conducted in several areas to select plants and identify species using the herbarium. The study used a systematic questionnaire and interviews to gather information on plant applications, cures for various conditions, medicine preparation, and potential risks. The information from each individual was recorded in the field notebook. Plant specimens were collected on several field trips, and herbarium sheets were created. Specimens were identified using literature and the Herbarium of the Department of Botany at the University of Rajasthan in Jaipur. Each plant was assigned an accession number from the latter. The current study demonstrates that Rajasthan is rich in Ethnoveterinary plant diversity.

Table 1. Ethnoveterinary plants of Rajasthan.

S. No.	Botanical name	Family	Common name	Parts used	Habit	Ethnoveterinary uses
1.	Abutilon indicum (Linn.) Sweet	Malvaceae	Kanghi	Leaves and seeds	Shrub	Use for relief in constipation, cure haematuria, diarrhoea, dysentery and use in foot and mouth disease.
2.	Acacia nilotica (Linn.) Del.. Sub sp. Indica (Benth)	Fabaceae	Babool	Thorns	Tree	Use in the easy removal of the placenta after delivery and for gastric problems.

	Brenan					
3.	Achyranthes aspera Linn.	Amaranthaceae	Adhijhara	Whole plant	Herb	Use to cure stomachaches, treat corneal opacity, and promote delivery.
4.	Aerva persica (Burm. f.) Merrill.	Amaranthaceae	Safed bui	Whole plant	Herb	Used to cure foot and mouth disease and digestive disorders.
7.	Azadirachta indica A. Juss	Meliaceae	Neem	Leaves	Tree	Use in foot and mouth disease, wounds and poisonous bites.
8.	Calotropis procera (Ait) R. Br.	Asclepiadaceae	Akra	Latex	Shrub	Used as an antidote for scorpion bites, for wound healing, and to treat khurpaka and gingivitis.
9.	Capparis decidua (Forssk) Edgew	Capparaceae	Ker	Stem	Shrub	Used in the healing of fractured bones.
10.	Celosia argentea Linn.	Amaranthaceae	Gorkha	Roots	Herb	Used in treating food poisoning and the emergence of placenta after delivery.
11.	Citrullus colocynthis (L.) Schard.	Cucurbitaceae	Gad-tumba	Roots	Herb	Used to cure constipation, abdominal pain and skin infection.
12.	Cocculus pendulus (Forst) Diels	Menispermaceae	Pilwan	Stem	Climber	Used to treat mastitis.
13.	Dalbergia sissoo Roxb.	Fabaceae	Tali	Leaves	Tree	Used to cure dysuria.

14.	Euphorbia hirta Linn.	Euphorbiaceae	Dudhi	Whole plant	Herb	Used to treat diarrhoea.
15.	Leucas aspera (Willd.) Link	Lamiaceae	Dargal	Whole plant	Herb	Used in treating fever, wounds and jaundice.
16.	Ricinus communis L.	Euphorbiaceae	Arandi	Seed oil	Shrub	Used to treat throat disease.
17.	Tecomella undulata (Sm.) Seem	Bignoniaceae	Rohida	Bark	Tree	Used to cure rashes on the skin.
18.	Tridax procumben s L.	Asteraceae	Kumra	Leaves	Herb	Used to treat diarrhoea and eye infections.
19.	Tribulus terrestris L.	Zygophyllacea e	Gokhru	Whole	Herb	Used to treat chronic cough and is beneficial in cases of strangury.
20.	Withania somnifera (L.) Dunal	Solanaceae	Ashwagan dha	Tubers	Shrub	Used to treat lumbago and fever.

Summary

The use of plants as medicine is gradually expanding among both rural and urban dwellers. According to the World Health Organisation (WHO), herbal medicines are used by 80% of the global population for primary healthcare purposes. Studies show that ethnic communities in the Rajasthan area rely on traditional knowledge of Ethnoveterinary plants for basic healthcare of the cattle of tribal people. Despite ongoing socio-cultural changes, residents nevertheless have a thorough understanding of plants and their applications.

Glossary

1. Ethnoveterinary: Systematic investigation and application of veterinary folk knowledge, theory, and practice

2. Indigenous: Traditional and local

Exercise: Answer the following questions.

1. What is Ethnoveterinary medicine?
2. Which plant is commonly used in Ethnoveterinary medicine to treat diarrhoea in livestock?

3. What is the significance of Ethnoveterinary plants in rural areas?
4. Which plant is used in Ethnoveterinary medicine to treat respiratory issues in livestock?
5. How are Ethnoveterinary plants typically used?
6. What is the benefit of using Ethnoveterinary plants?
7. What is the role of local communities in conserving Ethnoveterinary plants?
8. Which plant is used in Ethnoveterinary medicine to treat skin conditions in livestock?
9. How can Ethnoveterinary plants contribute to sustainable agriculture?
10. What are the challenges in promoting the use of Ethnoveterinary plants?

Test Paper

Chapter 7
Ethnoveterinary Practices of Rajasthan

Time Duration: 30 minutes **Maximum Marks: 25**

General Instructions:

1. The question paper consists of 12 questions and is divided into four sections, A, B, C and D.
2. All questions are compulsory.
3. Section A comprises MCQS, carrying one mark each.
4. Section B comprises VSAQS (very short answer questions) carrying three marks each.
5. Section C comprises SAQS (short questions) carrying three marks each.
6. Section D comprises long questions carrying six marks each.

SECTION-A

Exercise 1: Multiple choice questions. **(1*4=4)**

1. What is Ethnoveterinary medicine?
(a) Modern pharmaceuticals for animals
(b) Traditional knowledge of plants and animals for animal healthcare
(c) Laboratory-synthesised compounds for animals
(d) Industrial agricultural products for animals
2. Which plant is commonly used to treat ticks and fleas in livestock?
(a) Neem (Azadirachta indica)
(b) Turmeric (Curcuma longa)
(c) Ginger (Zingiber officinale)
(d) Garlic (Allium sativum)
3. What is the primary benefit of utilising Ethnoveterinary plants?
(a) Faster efficacy
(b) Lower cost
(c) Increased availability
(d) Reduced environmental impact
4. Which organisation promotes the use of Ethnoveterinary plants?
(a) World Health Organisation (WHO)
(b) Food and Agriculture Organisation (FAO)
(c) World Veterinary Association (WVA)
(d) International Union for Conservation of Nature (IUCN)

Answer: 1. (b) 2. (a) 3. (b) 4. (b)

SECTION-B

Exercise 2. Very Short answer type questions **(3*3=9)**

1. What are Ethnoveterinary plants?
2. Give examples of Ethnoveterinary plants.
3. Which regions of Rajasthan use Ethnoveterinary plants?

SECTION-C

Exercise 3. Short answer type questions **(3*2=6)**

1. What are the medical properties of plants used in Ethnoveterinary practice?
2. What are the threats to the conservation of Ethnoveterinary plants?

SECTION-D

Exercise 4. Long answer type questions **(6*1=6)**

1. How can ethnoveterinary practices be integrated into modern veterinary medicine?

Online Resources

	Online Practice Paper	Online Video Link
Link	https://forms.gle/Tu2w5WfpkkMJe8DC8	https://www.youtube.com/playlist?list=PLzlh_31VWEFqvJx-GwR3B2nkbq7wAo-dk
QR Code		

Chapter 8: Conservation Efforts in Rajasthan

In this chapter, you will learn about:

1. **Protection of Biodiversity**
2. **Conservation of Endangered Species**
3. **Sustainable Use of Natural Resources**
4. **Community-Based Conservation**

Conservation of plants and plant products is the part of consciousness and life style of the tribals because the most surprising and amazing traditions of tribals in India are linked anyway, with plants and their products and tradition are practiced in festivals, ceremonies, worship of god, worship trees and even the controlling and curing of the disease, the plants are in use from the immemorial time.
There are two ways of conservation adopted by the tribals.
1. Traditional/conventional method for the conservation of plants and their valuable parts.
2. Non-conventional/Indirect method for the conservation of plants

Plant conservation refers to the practices of protecting and preserving plant species, their habitats, and their ecosystems to ensure their survival for the use of future generations. Therefore, plant conservation is essential for the sustainable future. It is also necessary to consider both wild and cultivated forms. It can be achieved by following these ways: Sustainable agriculture: Plant crops must be planted and chosen in such a way that maximises the productivity while minimising negative impacts to the environment. It includes practices like crop rotation, organic farming, and reducing the use of pesticides. Forest management: Utilise the forest sustainably by striking a balance between resource extraction and conservation. This includes selective logging, reforestation and protection of endangered species. The management also leads to the viability of species within the ecosystem. Medicinal plant conservation: The conservation of wild medicinal plants, as well as other threatened species, can be addressed through both scientific techniques and social actions. They include discouraging the cutting down of indigenous trees and encouraging local people to plant fast-growing exotic and indigenous trees for domestic use, the inauguration of a National Tree Planting Day, and the creation of nature reserves. The Indian agency responsible for the conservation and sustainable use of medicinal plants is the Indian Council of Medical Research, New Delhi. Threats to plant diversity Include Habitat loss, fragmentation, and degradation. Habitat loss and associated fragmentation are the most significant single threats to plant diversity. Deforestation is the leading cause of habitat loss, fragmentation and degradation. It is caused to fulfil our requirement of dwelling, as the population is increasing at an alarming rate. People are shifting towards the nuclear family, which increases the demand for housing, transportation, and other necessities, such as invasive species. Invasive alien species are another potential threat to native plant diversity. A recent study revealed that more than 13,000 species, approximately 3.9% of the world's vascular plant flora, have become naturalised outside their native range due to human activities. The acclimatisation of invasive species leads to the extinction of native species. Ethnic groups have imposed restrictions and cultural inhibitions in the form of rites, rituals, and folk tales and lore to instil a sense of fear and conserve knowledge. The tribe has identified certain plants that are favourable to them, leading to prosperity based on their utility in protecting them. They worship them and feel cutting them off is a sin, and they punish anyone caught red-handed. Ficus religiosa (Pipali), Ficus. bengalensis (Bad), Adansonia digitata(Kalpvriksha), Musa paradisiaca (Keli), Phyllanthus emblica(Amla), Ocimum species (Tulsi) etc are among few examples of plants worshiped by people in the region Most of these plants are protected as some local deities are placed under these trees and sacred groves dedicated to deities of dimension ranging from a clump of few trees to vegetation spreading over several hectares are features of several tribal societies. Tribals celebrate different festivals, and it has been invariably noticed that all the above festivals are judiciously celebrated by the tribals with full enthusiasm and zeal by performing customary traditions that are related to different plants. Moreover, the conservation of biodiversity and the sustainable use of plant resources are another area that requires due attention. Surrounding their vicinity and these resources play an important role in their routine life. It is the need of the hour to focus immediate attention on plant conservation from the government and NGOs, with the help of local people, by creating rapid awareness among them. In situ conservation can be effectively achieved by establishing a system of protected areas, such as sanctuaries, National parks, and biosphere reserves. Ex–situ conservation involves the conservation of plants outside their natural habitats. Biotechnology and tissue culture would help preserve genetic material. The indigenous people and local communities are directly associated with the forestry operation. The protection and conservation of forest resources are not only desirable but also necessary for the nation's economic development. To protect and sustainably utilise forest resources, as well as validate and invalidate ethnic practices for the welfare of local communities, is essentially the need of the hour.

In India, religious values have played a crucial role in preserving the country's forests and environment. Many communities have established self-imposed rules and regulations regarding logging and hunting to conserve the natural environment. They earmark specific patches of land, which serve as an excellent refuge for native biodiversity, in the name of the local forest deity. The presiding deity is a local Hindu god, although some are based on Islamic and Buddhist origins, or folk religions. Due to a community-supported initiative, the local flora and fauna survive and thrive. These Patches of Forest are known as Sacred Groves and considered as "Sacred Natural Sites". These are the relic forest patches preserved in the name of religion and culture. They extend throughout the world from Asia, Africa, and Europe to America. In India, Sacred Groves are found in the Northeast Himalayan region, the Western and Eastern Ghats, the Coastal region, the Central Indian Plateau, and the Western Desert of Rajasthan.

Indian sacred groves have pre-Vedic origins and are associated with indigenous communities who believe in the divinity of nature and natural resources. There are more than one lakh such sacred groves in the country, which have different names in different states, such as Devrai, Devban, Devbhumi, Gompa, etc. In Rajasthan, they are called "Oran", derived from the Sanskrit word "Aranya", meaning forest. They are patches of jungle preserved in the name of local deities or saints. Sacred groves are known by various names in Rajasthan, including Deora, Malvan, Deorai, Rakhat Bani, and Devban. Rajasthan provided fascinating insights into tribal traditions of maintaining sacred groves.

Introduction

Sacred Groves are considered "Sacred Natural Sites" (IUCN) [1]. These are the relic forest patches preserved in the name of religion and culture. They extend from Asia, Africa, and Europe to America, mainly in Africa and Asia [2]. In India, Groves are present in the North-eastern Himalayan region, the Western and Eastern Ghats, the Coastal region, the Central Indian Plateau, and the Western Desert [3]. Indian sacred groves have a pre-Vedic origin. They are associated with indigenous and tribal communities, which believe in the divinity of nature and natural resources. The present thrust is on ecosystem functioning, ecological services and biodiversity conservation importance [4, 5]. There are more than a lakh sacred groves in the country, which have different names in various states, such as Devrai, Devban, Devbhumi, and Gompa. In Rajasthan, they are called Oran. Oran is derived from the Sanskrit word aranya, meaning forest. They are patches of jungle preserved in the name of local deities or sai. It is often believed that during shifting cultivation, a part of the forest is left undisturbed. Here, all the species found in the area are protected. These areas may have developed as sacred groves (Hazra, 1974, 1980; Gadgil & Vartak, 1976). Such sacred groves often protect watersheds and water sources. There is a popular theory that sacred groves, which protect a water source, may have originated from people's belief that a deity located near the grove yields water for agriculture. In a state like Rajasthan, where water is scarce for farming, animal husbandry and drinking purposes, it seems a reasonable factor. Forests in hills reduce runoff and help recharge groundwater. The water thus becomes available in the 'Bawdi' (step-well) or 'Kund' (pool) located within the sacred grove during the lean months. Water also brings minerals and fertilisers in rich quantities. It is then logical that such resources are protected and conserved by the people. People might have institutionalised these arrangements over time by attaching sacred value to them, making collective management easier and more long-lasting. Sacred groves are the result of a complex Ethnoscientific thinking among local communities (Pandey, 1996). People do not cut wood for personal use, and only dead and fallen trees are removed for use. The area under an Oran can vary from a few square meters to several hundred hectares.

According to the Forest Department, the protected areas of the forest are typically those where indigenous people have a local deity to worship. These orans are generally spread over a minimum of 5 hectares to 1500 hectares of land, and 200 trees per hectare should be the density of forest to classify them under sacred Groves. However, due to shrinking forest cover, areas of Orans spread over less than 5 hectares are also considered Sacred Groves in the Forest Department's work plan.

A unique aspect of Oran management is the presence of a mahatma, a Baba, or a priest with a socio-cultural role. The mahatma, like a watchman, resides in the Oran and keeps watch over the forest and its inhabitants, in exchange for introductory provisions from the community. He has been instrumental in fostering a reverence for nature and protecting the groves. People do not harm sacred groves mainly because of socio-religious traditions and fear of the unknown, believing that the presiding deity may harm those who cut or use an axe in a sacred grove. These beliefs might have strongly influenced the conservation of sacred groves.

Sacred groves in the Aravallis and Vindhyas can be classified into three major groups:

1. The first group consists of groves located near the village and close to a water source. Such groves are located at the top of small hillocks in the Aravallis of the Vindhyan tract, in Kota, Bundi, Baran, and Jhalawar, where people worship Bheruji, Bawsi, and Mataji. Ramgarh Mataji, Bijasan Mataji, and Sagas Ji Maharaj are examples of this type of Sacred grove, which were studied in this research programme.
2. The second group of groves is dedicated to Lord Mahadeo. Vegetation of the entire watershed or part of it is often protected as a grove. Large trees and water sources are the main characteristics of these groves. Dhundhla Mahadev Bundi, Gaipernath in Kota, Jharne ke Mahadev in Jhalawar are examples of this type of Sacred Grove. Water sources, such as open and step wells (Bawdi), may be seen. Sometimes, both groups can also be found in the same village.
3. The third type may be seen as a single tree. In the Kota forest range, several large trees of Ficus benghalensis are visible, and due to the development of aerial roots, these trees resemble a grove. Azadirachta indica trees surround Devnarayan ki Jhopri, and Mangifera indica trees surround Sitabari in Baran. The tradition of protecting Peepal, Gular and Bargad trees is also found. In the Hadoti region, certain castes are responsible for protecting specific species of plants, a practice known as their Dehari. People of a particular caste do not cut their Dehari and protect it by worshipping.
4. The Baug: Sacred Gardens. The Baug is an Ethno-garden planted near settlements for fruit, fodder, fuelwood, medicine, and shade. They are the backbone of indigenous methods for drought prevention, acquiring entitlements, and ensuring food security. The biodiversity consists of utility trees such as Mangifera indica, Madhuca latifolia, Feronia limonia, Syzygium cuminii, Tamarindus indica, and Phoenix, among others. Such baugs are present in Kota (inside the Darrah Wildlife Sanctuary), Bundi, Baran and Jhalawar districts in Rajasthan. Sacred gardens are cultivated counterparts of sacred groves. Every garden has a sacred place dedicated to a village deity. Green felling is banned by the community, only dead and fallen wood is removed by the owner.
5. Temple Forests: Temple forests are found in various countries, including India, China, Nepal, Thailand, and Indonesia. Temple forests are managed and maintained to serve the temple, encompassing economic, ecological, social, and religious functions. The Dadh Devi

temple in Kota, Rajasthan, India, is home to a large temple forest owned by the Temple Trust, as well as a sacred grove.

6. Sacred Corridors: Sacred Corridors are locally protected riverbanks that villagers have dedicated to Lord Shiva. Long stretches of Karai van or riparian forests are protected in several places along the river Chambal. In Kota Gaipernath and Garadia Mahadeo, sacred corridors along the river Chambal attract hundreds of visitors during the annual religious fair dedicated to Lord Shiva.

7. Another category of SGs includes those believed to be the abodes of ancestral spirits. Often these groves are usually burial grounds. Such groves have been reported from several places in Hadoti. Nau Gaje Peer in Bandha (Ramganjmandi) and Peer Baba Ki Majar in Shergarh Sanctuary are examples of burial grounds as Sacred Groves.

The Hindu tradition categorises forests into three types: Tapovan, Mahavan, and Sreevan. Tapovan is a forest associated with penance (Tapas) and is inhabited by saints and rishis. Mahavan refers to the grand natural forests. Tapovan and Mahavan are considered to be a Raksha ("sanctuary") for flora and fauna, as ordinary human beings are not allowed to enter these forests. Sreevan, which means "forests of prosperity", consists of dense forests and groves. From the former, people would collect dry wood, leaves, forest produce and a limited amount of timber, though the natural ecosystem would not be unnecessarily disturbed. Groves were considered spaces of forests from where harvesting could be done. Sometimes, specific trees, such as mango trees, can be planted and nurtured here, as seen in the Sitabari Temple forest. Groves were associated with religious rites, festivals and recreation. Typical recreational activities associated with these groves included annual Fairs and festival celebrations. In the villages, a Panchavati, or a cluster of five trees representing the forest, was maintained. These trees represented the five elements of Earth, Water, Fire, Air and Space.

Planting and nurturing of trees has been a highly evolved practice in ancient India. Vrukshayurveda, the science of plant life, is also a 10th-century treatise of that title on the subject, ascribed to Surapala, which dealt with various species of trees and their growth.

Plants worshipped and protected by particular castes:
In the Kota forest region, several large trees of Ficus benghalensis, which are over 200 years old, are revered as sacred groves. Due to the development of aerial roots, these trees resemble groves and cover large areas. The tradition of protecting Peepal, Gular and Bargad trees is also seen. In the Hadoti region, certain castes are responsible for protecting specific species of plants, a practice known as their Dehari. People of other castes do not cut their Dehari and protect it by worshipping these trees. A well-known example of such protection is found in the arid region of Rajasthan, such as Jodhpur and Pali, where the Bisnois protect trees of Prosopis cineraria (Khejri). Similar cases reported in South East Rajasthan are as follows:

Caste	**Plant Species**

Royal Family of Kota	Ashapala
Gulgama	Ber
Sagitara	Jowar
Solanki Rajput	Palash
Bagtariya	Haldi
Gurjar	Neem
Bagdi	Mango
Sakwadia	Mango
Sahariya	Mango
Taroliya	Banana, Kair

Tulsi, Bargad, Goolar, Amla, and Peepal are auspicious plants worshipped by all Hindu castes, who take great care to conserve them.

Uses of Sacred Groves:

It can be further divided into two broad categories:

Traditional uses: One of the most significant traditional uses of sacred groves was that they served as repositories for various plants used as medicinal treatments in the traditional system of medicine. Other uses involved sources of renewable resources, such as fruits and honey. However, in most sacred groves it was taboo to hunt or cut down wood. The vegetation cover helps reduce soil erosion and prevent desertification, which is particularly beneficial in Rajasthan. The groves are often associated with ponds and streams, and meet the water requirements of local communities. They also sometimes help recharge aquifers. A large number of distinct local art forms and folk traditions are associated with the deities of sacred groves and are an important cultural aspect closely tied to sacred traditions, ritualistic dances and dramatisations based on the local deities that protect the groves. Sacred groves also serve as meeting places for the community to discuss socio-religious and economic issues and resolve their grievances.

Modern uses: In modern times, sacred groves have become biodiversity hotspots, as various species seek refuge in the areas due to progressive habitat destruction and hunting. Sacred groves often contain plant and animal species that have become extinct in neighbouring areas. They therefore harbour great genetic diversity. Additionally, sacred groves in urban landscapes also serve as "lungs" for the city, providing much-needed vegetation cover.

Threats to the Sacred Groves:

Threats to the groves include urbanisation, overexploitation of resources (such as overgrazing and excessive firewood collection), Mining, encroachment, and other depletive and environmentally destructive practices, which also pose a key threat to sacred groves. Other threats to the sacred groves include invasion by invasive species, such as the weeds Chromolaena odorata, Lantana camara, and Prosopis juliflora. Unfortunately, these areas have not even been designated as forest lands; hence, effective legislation is not applied in cases of offenders. Saranas are fairly degraded landscapes with over-mature Shorea robusta trees, exhibiting minimal regeneration due to open grazing, non-timber forest product collection, and various other biotic pressures (Patnaik & Pandey, 1998). In many places, local folk deities have been, and continue to be, replaced with Hindu gods and goddesses.

Summary

The institution of SGs is very ancient in our country. Access to and interference with SGs have been culturally restricted, thereby reducing the human impact on the harvesting of natural resources. The consequence of such a restriction has been that SGs have evolved as important reservoirs of biodiversity. SGs have become changed habitats housing a variety of genetic pools. Ecologically valuable species, such as Albizia lebbeck and Ficus glomerata, which conserve a high amount of nitrogen, phosphorus, magnesium, and calcium in their leaves, are primarily found in the semi-arid zones (SGs) of the southeast region of Rajasthan. In' orans of Western Rajasthan, the khejari (Prosopis cineraria) is a keystone species, similarly Terminalia Arjuna, Ficus bengalensis in orans of South –east Rajasthan inseparably linked to the survival of many other species, and occupies a special position in the Rajasthani culture was also seen in the present course of study. Presently, even as the Government spends considerable amounts on protecting reserved forests and wildlife sanctuaries, it is unable to provide the protection that existed in ancient sacred groves, where indigenous communities took direct charge of their own environmental needs. To ensure the survival of such areas in their original condition, we have to promote our traditional and indigenous system of conservation in the present scenario, which is the need of the hour. To conclude, considerable consideration has been given to the scope and importance of Ethnobotanical studies. Moreover, there is a need to conduct detailed research on plants of medicinal value, as well as plants of medicinal importance that have yet to be studied in various aspects, including their biochemical properties. It has relevance in the conservation of genetic resources and helps to search for new sources of drugs, food, fodder and other life-supporting species found in nature.

Glossary

1. Abiotic: Non-living factors that influence plant growth, function and survival (for example, soil, water, amount of sunlight, nutrients, etc.). Acclimate: Harden off a plant so that it can survive in low-humidity conditions. Bioinformatics: An interdisciplinary field of science that combines biology, computer science, mathematics, and statistics to analyse and interpret biological data.

2. Biotic: Living factors that influence plant growth, function and survival (for example, predators, fungi, pollinators).

3. Conservation Status: A formal designation of endangered or threatened status at a state, federal, or international level. Cryopreservation: Preserving tissues by storing them in liquid nitrogen at temperatures below - 130°C.

4. DNA Banking: Long-term storage of an individual's genetic material. In plants, leaves are a common tissue that can be stored for future use. Ex situ: Offsite, away from the wild population, typically referring to collections held in a nursery or botanical garden. Germplasm: Living seeds or tissues from which plants can be grown.

5. Herbarium: It is the collection of completely dried, properly pressed, perfectly mounted plant specimens on the herbarium sheets.

6. Sacred Groves- Sacred groves, sacred woods, or sacred forests are groves of trees that have special religious importance within a particular culture.

Exercise: Answer the following questions.

1. What is a sacred grove?
2. What is the significance of sacred groves in biodiversity conservation?
3. What is meant by in situ conservation?
4. What is the role of local communities in conserving sacred groves?
5. How do sacred groves contribute to ecosystem services?
6. What is the significance of sacred groves in maintaining cultural heritage?
7. What is ex-situ conservation?
8. How can conservation efforts be strengthened in sacred groves?
9. What is the importance of documenting traditional knowledge associated with sacred groves?
10. How can sacred groves be integrated into larger conservation efforts?

Test Paper

Chapter 8
Conservation Efforts in Rajasthan

Time Duration: 30 minutes **Maximum Marks: 25**

General Instructions:

1. The Spaper consists of 12 questions and is divided into four sections, A, B, C and D.
2. All questions are compulsory.
3. Section A comprises MCQS, carrying one mark each.
4. Section B comprises VSAQS (very short answer questions) carrying three marks each.
5. Section C comprises SAQS (short questions) carrying three marks each.
6. Section D comprises long questions carrying six marks each.

SECTION-A

Exercise 1: Multiple choice questions: **(1*4=2)**

1. What is the primary goal of conservation efforts in Rajasthan?
(a) To promote eco-tourism
(b) To protect biodiversity and ecosystems
(c) To support agricultural development
(d) To reduce poverty
2. Which of the following is a protected area in Rajasthan?
(a) Ranthambore National Park
(b) Sariska Tiger Reserve
(c) Keoladeo National Park
(d) All of the above
3. What is the primary objective of community-based conservation in Rajasthan?
(a) To involve local communities in conservation efforts
(b) To promote sustainable livelihoods
(c) To protect wildlife habitats
(d) All of the above
4. Which organisation is involved in conservation efforts in Rajasthan?
(a) Rajasthan Forest Department
(b) Wildlife Institute of India
(c) World Wildlife Fund (WWF)
(d) All of the above

Answer 1. (b) 2. (d) 3. (d) 4. (d)

SECTION-B

Exercise 2: Very Short answers type questions **(3*3=9)**

1. What is the main challenge conservation efforts in Rajasthan face?
2. What is the role of the Rajasthan State Biodiversity Board in conservation efforts?
3. What is the main challenge of community-based conservation initiatives in Rajasthan?

SECTION-C

Exercise 3: Short answer type questions **(3*2=6)**

1. What is the primary goal of conservation efforts in Rajasthan?
2. What is meant by community-based conservation?

SECTION-D

Exercise 4: Long answers type questions **(6*1=6)**

1. Discuss in detail the various conservation efforts in Rajasthan to protect the Ethnomedicinal plants.

Online Resources

	Online Practice Paper	Online Video Link
Link	https://forms.gle/zYfBpbkwH4LuEBxL7	https://www.youtube.com/playlist?list=PLzlh_31VWEFpYMNcC6w3TaM4Ozvz5Py1E
QR Code		

Chapter 9: Conclusion

In this chapter, you will learn about:

1. **Problems occurred during the Ethnobotanical survey**
2. **Strategies for overcoming the problems**

The tribal people residing in various parts of Rajasthan are a treasure trove of indigenous knowledge about the Ethnomedicinal plants of the region, which are used to cure their health ailments as well as those of their cattle. These individuals are very reluctant to share their information with the unknown; therefore, a rapport must be established with them to gather this information. As the young generation is not interested in this information and has shifted to urban sites, it will not be passed on to future generations. It will be lost forever if not documented. During the survey to collect this information from the tribal people, various problems were faced, which are mentioned -

1. During the Ethnobotanical survey, people are often reluctant to share information about their traditional knowledge of plants and their benefits.
2. As indigenous traditional knowledge is not documented and transferred from generation to generation verbally, information loss occurs.
3. Difficulty in finding the right individuals who possess Indigenous knowledge, as the local people may be prejudiced against officials dealing with land records or income tax, and thus may be wary of researchers as well.
4. Researchers may be unable to understand the dialectical nomenclature used by local people to describe plant species.
5. Traditional healers and older adults possess sufficient knowledge of plants, but they are concerned about the potential overexposure of these plants for commercial applications by the pharmaceutical and drug industry.
6. Due to modernisation, the tribes have become more civilised and are thus no longer willing to reside in villages; instead, they are moving towards cities, which often involves ignoring their traditional values, customs, and beliefs.
7. Sometimes, it is also challenging to identify plants correctly by their local names, and confusion arises over the Latin names of the plants during documentation. Various strategies have been tested to overcome this problem, which are as discussed-

Strategies to overcome the problems

1. During an Ethnobotanical survey, the informant should be informed that the indigenous knowledge provided by the tribes will not lead to over-exploitation of plants.
2. Irrespective of modernisation, the importance of medicinal plants and their uses in developing countries must be taken seriously. It can be adequately achieved using bioinformatics tools and techniques.

3. Conducting group meetings to promote awareness and motivate local people to share their knowledge.
4. Conducting location-specific group meetings, demonstrations, and home visits to address local people's problems and priorities (e.g., insects, disease) could encourage local people to reciprocate by sharing their knowledge.
5. Extension teaching methods and aids help educate local people, and appropriate counselling ensures the availability of the required knowledge.
6. Arrange scouting competitions for local people (women, farmers, children, etc.), and select winners who should be awarded rewards or incentives that can motivate them to share their Indigenous traditional knowledge.
7. The socioeconomic standard of the tribal people should be improved, as it reduces their move towards the city for livelihood.

Proper documentation, identification, and conservation of important traditional medicinal plants should enhance research capabilities in this field. Government laws and policies will facilitate the implementation of necessary conservation practices.

Positive Implications of Ethnobotany

1. **Preservation of Traditional Knowledge:** Ethnobotany helps preserve the traditional knowledge of plants and their uses, which is essential for Rajasthan's cultural heritage.
2. **Conservation of Biodiversity:** Conservation efforts in Rajasthan help protect the state's rich biodiversity, including medicinal plants, which are essential for human health.
3. **Sustainable Livelihoods:** Ethnobotany and conservation can provide sustainable livelihoods for local communities, such as through eco-tourism and sustainable harvesting of plant resources.
4. **Cultural Revitalisation: Ethnobotany can help revitalise** cultural practices and traditions related to plant use, fostering community pride and identity.
5. **Improved Human Health:** The conservation of medicinal plants ensures their continued availability for traditional medicine, thereby improving human health and well-being.

Negative Implications of Ethnobotany

1. **Over-Exploitation of Plant Resources:** The increased demand for plant-based products can lead to the over-exploitation of plant resources, thereby threatening their sustainability.
2. **Loss of Traditional Knowledge:** Without proper documentation and preservation, traditional knowledge of plants and their uses can be lost, especially among younger generations.
3. **Cultural Appropriation:** Ethnobotany can lead to cultural appropriation, where traditional knowledge and practices are adopted without proper understanding, respect, or compensation.
4. **Conflict over Resource Use:** Conservation efforts can lead to conflicts over resource use, especially between local communities and external stakeholders.
5. **Climate Change Impacts:** Climate change can impact plant populations and ecosystems, making conservation efforts more challenging.

Future Directions

1. **Integrate Traditional Knowledge with Modern Science:** Combine traditional knowledge with modern scientific research to develop effective conservation strategies.
2. **Promote Sustainable Livelihoods:** Support sustainable livelihoods for local communities, such as through eco-tourism and sustainable harvesting of plant resources.
3. **Develop Community-Led Conservation Initiatives:** Empower local communities to lead conservation initiatives, ensuring their involvement and benefit-sharing.
4. **Address Climate Change Impacts:** Develop strategies to mitigate the impacts of climate change on plant populations and ecosystems.
5. **Foster Collaboration and Partnerships:** Encourage collaboration and partnerships among stakeholders, including local communities, researchers, and policymakers, to achieve effective conservation outcomes.

Recommendations

Some alternative arrangements should be implemented to address the economic needs of tribal and rural populations, enabling them to transition away from their traditional economic activities and conserve plant biodiversity. The following suggestions may be discussed in this regard.

1. The reorientation of the forestry programme should be taken up by the forest department and the local tribal people, under the "Joint forest management programme" to maintain their heritage. It may be necessary to upgrade these skills and provide them with a broader understanding so that they can appreciate their role in biodiversity conservation.

2. There is an urgent need for the conservation of genetic resources by declaring the sensitive areas as "Biological Diversity Research Centre". This will go a long way towards safeguarding the plant resources and the healthy ecological environment of that particular area.

3. The Forest Department of the Rajasthan State Government now undertakes eco-restoration and eco-development. It is recommended that the work be expedited at a faster rate.

4. Non-timber forest products (NTFPs) are an essential category of commodities in the local and National trade, regulated by businesses. The economy based on these products is largely invisible and regional, and it is part of an informal economy. Thus, it is not captured by the national accounting system. This should be utilised as an instrument for the economic development of Rajasthan's tribal and rural areas.

5. Women-headed community-based biodiversity conservation programmes are highly effective and successful. They are highly involved in art and craft, bamboo work, lac collection, oil extraction and food preservation. Total employment and income generated by forest-based small-scale enterprises should be encouraged through the Women's Entrepreneur Programme. Small-scale enterprises based on local non-timber forest products (NTFPs) for women should be prioritised, and incentives should be provided to bring them on par with the international market.

6. It is observed that the people in the village are capable of fulfilling a significant share of their fuel wood demand from their land. Therefore, it is imperative to develop a strategy for undertaking large-scale energy plantation programs by establishing woodlots of suitable species on waste lands and unused agricultural lands. Fuel wood should be converted into value-added products, such as charcoal, ethanol, and menthol, which can serve as an alternative energy source. There is a need to explore alternative sources of energy, such as solar and wind energy, which are often found to be viable options for addressing energy requirements.

7. There is a need for more planting of trees, especially endangered and threatened species, which are becoming locally extinct due to forest management practices that are also urgently needed. The government should establish enabling conditions for forest (wood) industries to provide a market for timber and offer forest owners incentives to utilise alternative products that offer few or no carbon benefits.

8. Promote the recyclability and recovery of forest products, as well as the responsible use of materials. The public should be encouraged to adopt sustainable and highly efficient practices for producing and utilising all types of biomass, thereby minimising competition for the land required to deliver food, energy, and industrial raw materials.

9. India is a rapidly growing economy in the world, with per capita Carbon dioxide (a greenhouse gas) emissions of only 1.1 tons. Hence, the forest and its allied sectors need to be critically emphasised in both the sustainable development and conservation of the country's resource wealth.

10. The most urgent need is to design and implement monitoring programs for plant species that still require large-scale collection from the wild. The harvesting of natural resources must be carried out sustainably, in the interest of long-term healthcare maintenance.

11. The role of local communities in conserving nature and plants, particularly as sacred groves, should be acknowledged. The government should give authority to a committee consisting of representatives nominated by the Village Panchayat or Gram Sabha and Forest officials. This committee will be responsible for managing and conserving the community. This will be a helpful tool to create a strong bond between forest dwellers and forest officials in the study area.

12. It is observed that shifting cultivation (Jhum cultivation) is a traditional farming practice that has come under severe criticism from nearly all quarters as an uneconomic, wasteful, and inefficient system; thus, there is a felt need to address it without further delay. Dairy agroforestry and agro-based small-scale industries can provide a regular income and employment to the Jhumias, while also reducing pressure on Jhum cultivation.

13. Invasive plants are the second most significant threat to natural ecosystems, following the destruction of natural habitats by human activities. Medicinal plants can provide a substantial source of income for tribal people through the sale of wild-harvested material. These people are usually landless and belong to economically marginalised sections of society. Medicinal plants hold an essential place in tribal lifestyles, which can be advantageous for conservation efforts, given that this acknowledgement is well-rooted in culture. The depletion of natural vegetation is due to the weakening of customary laws that traditionally have regulated the use of natural resources. Such laws have often proved to be easily undermined by modern socioeconomic forces.

14. During the surveys conducted in the study area, the findings revealed that every year, the total amount of human knowledge about the types, distribution, ecology, methods of management, and preparation of herbal formulations is vanishing at a drastic rate. For conserving this knowledge, authorities need to identify the conditions in the region that are most favourable for releasing the potential offered by medicinal plants to promote sustainable development. Locals should revitalise their knowledge of locally available medicinal plants.

Summary

Ethnobotanical research can be considered a connecting link between medicine, past and present, in terms of plants and traditional societies. It is a crucial tool in the development of pharmaceuticals and the pharmaceutical industry. Complementary therapy for dermatological problems and treatment remains the primary option for millions of people in the Indian subcontinent. These characteristics of Ethnoveterinary knowledge articulate its vitality and the flexibility of many of its knowledge spheres. The majority of this vast knowledge trove is confined to the domain of older people, rather than the younger generation in society. This is a very alarming concern identified in the context of the sustainability of the Ethnoveterinary medicine knowledge system. In the future, Ethnobotany may play an increasingly important role in sustainable development and biodiversity conservation. Thus, breaking the forbidden chains between indigenous traditional knowledge and development. However, to cope with the increasing demand for medicinal and aromatic plants worldwide, particularly in the pharmaceutical and nutraceutical industries, our country must maintain a balance between availability and supply. Hence, it is a need of the hour to develop appropriate technologies to fulfil present and future demands in the most sustainable and balanced ways. One effective method in this regard is to implement commercial cultivation.

Glossary

1. Ethnobotany: It is the study of how people of a particular culture and region make use of indigenous (native) plants.

2. Conservation: It is the careful maintenance and upkeep of a natural resource to prevent it from disappearing

3. Preservation: The activity or process of keeping something valued alive, intact, or free from damage or decay.
4. Indigenous: Local or of an area
5. TK: Traditional Knowledge

Exercise: Answer the following questions.

1. What is the significance of Ethnobotany in modern medicine?
2. How can Ethnobotany contribute to biodiversity conservation?
3. What is the role of Ethnobotany in promoting cultural heritage?
4. How can Ethnobotany inform sustainable development practices?
5. What are the potential economic implications of Ethnobotany?
6. How can Ethnobotany contribute to mitigating and adapting to climate change?
7. What are the potential social implications of Ethnobotany?
8. How can Ethnobotany inform policy and decision-making?
9. What are the potential challenges and limitations of Ethnobotany?
10. How can Ethnobotany contribute to achieving the United Nations' Sustainable Development Goals (SDGs)?

Test Paper

Chapter 9
Conclusion: Summary of key findings and implications for Ethnobotany and conservation in Rajasthan

Time Duration: 30 Minutes **Maximum Marks: 25**

General Instructions:

1. The paper consists of 12 questions and is divided into four sections, A, B, C and D.
2. All questions are compulsory.
3. Section A comprises MCQS, carrying one mark each
4. Section B comprises VSAQS (concise answer questions), carrying three marks each.
5. Section C comprises SAQS (short questions) carrying three marks each.
6. Section D comprises long questions carrying six marks each.

SECTION-A

Exercise 1. Multiple choice questions: **(1*4=4)**

1. What is the significance of Ethnobotanical studies in Rajasthan?
(a) To document traditional knowledge of plants
(b) To promote the conservation of plant species
(c) To develop new medicines from plants
(d) All of the above
2. Which of the following is a threat to plant diversity in Rajasthan?
(a) Over-grazing
(b) Deforestation
(c) Climate change
(d) All of the above
3. What is the role of local communities in the conservation of plant species in Rajasthan?
(a) To provide information on the traditional uses of plants
(b) To participate in conservation efforts
(c) To benefit from the sustainable use of plant resources
(d) All of the above
4. Which of the following is an initiative for the conservation of plant species in Rajasthan?
(a) Rajasthan State Biodiversity Board
(b) Forest Department of Rajasthan
(c) World Wildlife Fund (WWF) - India
(d) All the above

Answer 1. (d) 2. (d) 3. (d) 4. (d)

SECTION-B

Exercise 2: Very Short answers type questions **(3*3=9)**

1. What are the different factors that lead to the loss of plant biodiversity?
2. Explain with examples the role of Ethnobotany in the conservation of plants.
3. What is the significance of Ethnobotanical research in promoting the conservation of plant species in Rajasthan?

SECTION-C

Exercise 3: Short answer type questions **(3*2=6)**

1. Differentiate between Ethnobotany and Economic botany.
2. What are the significant roles of the herbarium in Ethnobotanical studies?

SECTION-D

Exercise 4: Long answers type question **(6*1=6)**

1. Explain different conservation strategies for the conservation of plant resources.

Online Resources

	Online Practice Paper	Online Video Link
Link	https://forms.gle/fdVPgsrdwWyiBUmA8	https://www.youtube.com/playlist?list=PLzlh_31VWEFoVSvakbLiYnwbuFKP0rSyw
QR Code		

Ethnobotany and Traditional Knowledge Preservation Videos Playlist-

Link: https://www.youtube.com/playlist?list=PLzlh_31VWEFpW-BHZZbi7weo6JG_o6WIi

QR Code: